D1807982

Fats, Oils and Waxes

to
Sheila, Anne and Wendy
for their patience and encouragement

Distillation of essential oils

Preface

The rapid and often revolut
since World War II have ma
pressed teacher to keep abrea
school and industry. In add
novel materials forged by the
related problems of scientific
use of knowledge, is unlikely
approach to the subject. In t
and socio-economic implicatio
cal technology, the present se
relevant background materi
changing patterns of industry
continuous production and
facture. An attempt has also
of substances are related to th
can be modified by molecular

Two complementary volu
Experiments, contain suggesti
survey of modern preparative

Although the text has been
science teachers it should also
and colleges as well as senior
ing in mind the wide spectrum
inevitable that certain errors
every effort has been made to
the chosen topics.

Chemistry in Industry

Fats, Oils and Waxes

Peter Tooley BSc MSc ARIC MIBiol

Head of Department of Chemistry
St Osyth's College, Clacton

John Murray Albemarle Street London

Printed in Great Br
London, Beccles an

0 7195 2027 4 boa
0 7195 2028 2 lim

Acknowledgments

It is a pleasure to acknowledge the ready help of many colleagues and friends in industry and Government departments who provided valuable advice, statistics and other material. I have also had to draw upon the published works of many authors who are too numerous to mention individually, but to whom my thanks are also due.

I am grateful for the help afforded by the staff of John Murray who have successfully steered the book through the hazards of production. Mention must also be made of the loyal assistance of my senior technician, Miss Sandra Fairbairn, who has not only deciphered and typed the MS but given freely of her time and technical expertise.

Thanks are due to the following who have kindly permitted the reproduction of copyright photographs: pages 2, 20, 39 (*top* and *bottom*), 73, 75 (*top* and *bottom*), 77 (*top* and *bottom*), 93, 168, Unilever Ltd; 21 (*top* and *bottom*), 26 (*top* and *bottom*), Rose, Downs & Thompson Ltd; 35, from *Historisches Bildarchiv-Handke*; 47 (*top* and *bottom*), Imperial Chemical Industries Ltd (Paints Division); 54, 95, Shell Petroleum Co. Ltd; 59, from *L'Art du Savonnier*; 103, Johnsons Wax; 113, 122 (*top* and *bottom*), Yardleys; 120, Apex Construction Ltd; 147, 149 (*top* and *bottom*), 155, Bush, Boake Allen Ltd; 171, US Embassy, London.

Thanks are also due to the following for permission to base diagrams on material from their publications: figures 2.1, 2.2, 2.3, 2.7, 3.7, 3.8, 3.9, 3.10, 3.11, 3.13, 4.2, 5.2, 5.3, 5.4, Unilever Ltd; 3.12, Shell Petroleum Co. Ltd; table on page 62, *Chemicals from Petroleum* by Waddams, John Murray Ltd (1967).

The cover photo is reproduced by courtesy of Albright & Wilson Ltd.

Contents

Chapter 1

Introduction

'It is an excellent thing to shew the diversity of ways to make Oyl. That if Olives should ever be scarce, yet we might know how to draw Oyl from many kinds of fruits and seeds.'

Natural Magick—Book 4, John Baptista Porta, 1658

At the present time the human race uses an estimated 40 million tonnes a year of fats and oils derived from both animal and vegetable sources. This reflects both their nutritional and industrial importance, and yet curiously the fats and waxes have remained for many years a neglected chapter in organic chemistry. More recently, however, interest has been stimulated by the advances which have been made in the manufacture of such materials as detergents, polishes, cosmetics and edible fat blends. In addition the unravelling of the complex chemistry of the essential oils has led to the development of technical perfumes and synthetic flavouring agents.

The true fats (glycerides) belong to a large group of natural organic polar compounds known as lipids, which also includes the waxes, lecithins and phospholipids. These are widely distributed in foods and are of great nutritional value. Fats provide fatty acids which are necessary to animal metabolism and represent a concentrated reserve of energy, yielding about 37 kJ (9 Calories) per gramme. This compares with corresponding values of 16·5 kJ (4 Calories) and 22·75 kJ (5·5 Calories) per gramme for carbohydrate and protein respectively. Although animal foods contribute a considerable proportion of our dietary fats some plant materials—especially seeds and nuts—are rich in lipids. Thus soya bean, groundnut, cottonseed, sunflower seed, coconut and palm kernel are all important sources of edible oils. Certain fruits such as the avocado and olive also have an unusually high fat content.

Animal fat is mainly located in adipose tissue cells. These are distended with oily droplets which solidify on death in warm-blooded

Palm kernels from which palm oil is extracted

Source	Fat content %	Source	Fat content %
Pecans	73	Safflower	30
Copra	69	Avocado	20
Walnuts	64	Cottonseed	20
Castor seed	50	Olive	19
Palm kernel	50	Soya	18
Sesame seed	50	Oatmeal	7·5
Groundnuts	45	Yam	4
Rapeseed	40	Maize	2
Linseed	36	Barley	1

animals. In the human body, fat forms about 12% of its total weight. At least half of this fat forms a protective heat insulating subcutaneous layer, which is part of the depot fat available to the body as a food. Smaller amounts are to be found in the intermuscular connective tissue, bones, and as a protective layer around nervous tissue, kidneys, heart and other organs. Fats and oils derived from water-living animals usually contain a greater variety of fatty acids than those obtained from land animals and plants. A substantial proportion of these fatty acids are unsaturated, the marine oils being mainly in the C_{20}–C_{22} range and the fresh water oils from C_{16}–C_{18}. Commercial fats and oils consist almost entirely of true fats which are the glyceryl esters of fatty acids. The fatty acids are so-called because many were first discovered as the result of hydrolysing the glyceryl esters occurring in fats. As glycerol is a trihydric alcohol it can react with one, two or three molecules of monobasic fatty acids to produce monoglycerides, diglycerides or triglycerides respectively. Only triglycerides are found naturally in undecomposed fatty material. A simple triglyceride contains only one type of fatty acid radical. Thus an important component of olive oil is triolein which is the triglyceride of oleic acid. The majority of naturally occurring triglycerides are mixed, however, containing two or three different fatty acid radicals. The structure of a triglyceride is often represented by the initials of its component fatty acid radicals. In this way 2-oleo 1,3-distearin would be identified as SOS.

Most fats are themselves mixtures of triglycerides and the proportions of these which are present determine their physical charac-

$$CH_2OH$$
$$CHOH$$
$$CH_2OH$$

glycerol

$$CH_2OCO(CH_2)_{16}CH_3$$
$$CHOCO(CH_2)_7CH=CH(CH_2)_7CH_3$$
$$CH_2OCO(CH_2)_{16}CH_3$$

2-oleo 1,3-distearin (SOS)
(a mixed triglyceride)

teristics. The oils, which are fats with low melting points, contain predominantly unsaturated fatty acids. Castor oil, for example, contains about 90% unsaturated ricinoleic acid. The presence of double bonds in the fatty acid chain permits the formation of isomers, which also affects the melting point. For example, (*cis*) oleic acid has a melting point of 16·3 °C while its isomeric form (*trans*) elaidic acid has a melting point of 43·7 °C.

Another factor which determines whether a fat is a solid or an oil is the molecular weight of the fatty acids involved. Thus hydrolysis of coconut oil produces a mixture including saturated fatty acids with low molecular weights, whereas the hydrolysate of beef 'dripping' yields long chain saturated fatty acids such as stearic acid and palmitic acid.

	Saturated fatty acids				Unsaturated fatty acids			
	My-ristic	Pal-mitic	Stearic	Arachidic	Oleic	Lin-oleic	Butyric	Caproic
Corn oil	—	5	2	2	36	55	—	—
Olive oil	1	8	2	1	80	8	—	—
Lard	1	27	9	—	55	6	—	—
Beef drip-ping	2	33	14	—	48	3	—	—
Butter	10	23	11	—	31	3	10	2

Approximate percentage distribution of a range of fatty acids in some common edible fats

When glycerides solidify they exhibit differences in their crystalline packing structure (polymorphism). This is due to the inability of the atoms to rotate around single bonds as in the liquid state and a consequent 'freezing' of the molecular form as solidification occurs. Three principal forms have been identified by X-ray diffraction techniques, an unstable alpha form produced by rapid cooling and two relatively stable beta forms (beta—β, and beta prime—β'). These molecular patterns in solid fats represent different orientations of the fatty acid chains with relation to the plane of the glyceride molecule. This is thought to be shaped like a tuning fork. Interconversion of one

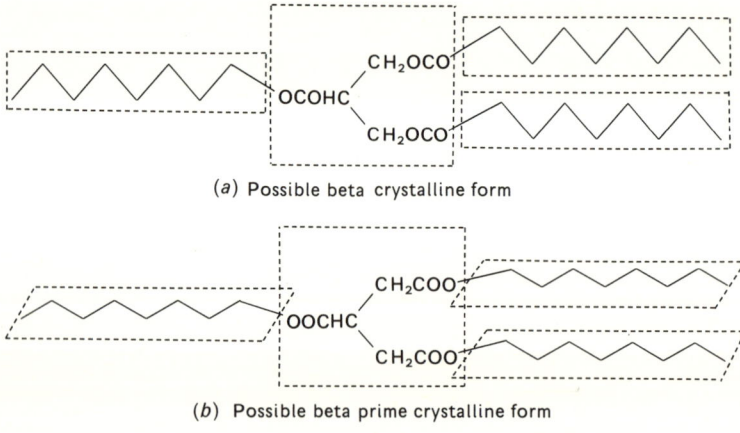

(*a*) Possible beta crystalline form

(*b*) Possible beta prime crystalline form

(∿ represents a fatty acid chain)

crystalline form into another can be followed by noting the small associated heat changes which occur during cooling (differential thermal analysis).

When pure, the glycerides and the fatty acids are colourless and usually almost odourless. The colour and smell of fats are due to the presence of small quantities of non-fatty materials. Thus the yellowish colour of some fats originates from fat soluble pigments such as carotenoids, and the characteristic smell of palm oil is largely due to small quantities of β-ionone.

The fatty acids which are found as esters of glycerol and other alcohols in the fats and waxes are usually straight chain compounds with an even number of carbon atoms in the molecule. An interesting and isolated exception is iso-valeric acid which is a C_5 acid found in the hydrolysate of dolphin oil. In addition to having an odd number of carbon atoms, iso-valeric acid has a branched chain. Some of the waxes found in bacilli also have branched chain fatty acids.

$$CH_3(CH_2)_7CH(CH_2)_8COOH$$
$$CH_3$$

tuberculostearic acid
(found in
Mycobacterium tuberculosis)

$$CH_3$$
$$CHCH_2COOH$$
$$CH_3$$

iso-valeric acid

Three fatty acids of plant origin (gorlic, hydnocarpic and chaulmoogric acids) have a hydrocarbon chain which terminates in a cyclopentene ring. Since these acids possess an asymmetric carbon atom they are able to form optical isomers.

$$HC=CH$$
$$CH(CH_2)_{10}COOH$$
$$H_2C-CH_2$$

hydnocarpic acid

$$HC=CH$$
$$CH(CH_2)_{12}COOH$$
$$H_2C-CH_2$$

chaulmoogric acid

The lower fatty acids are soluble in water but as the chain length increases solubility falls rapidly. The higher members of the series are insoluble in water although readily soluble in the usual organic solvents. All the fats are insoluble in water, and, with the exception of castor oil, are only sparingly soluble in ethanol. The ready solubility of fats in most organic solvents is utilized in solvent extraction techniques.

The double bonds present in the hydrocarbon chain of the unsaturated fatty acids enable stereoisomeric forms to exist. This is due to the restriction on free rotation around the double bonds which allows the formation of different molecular structures. Thus oleic acid and elaidic acid are geometrical isomers, the naturally occurring oleic acid having both hydrocarbon chains on the same side of the double bond (*cis* form), elaidic acid having them on opposite sides (*trans* form). Only the *cis* isomers occur naturally.

oleic acid (*cis* form)   elaidic acid (*trans* form)

This may be shown in diagrammatic form for convenience, the thick lines representing hydrocarbon chains.

oleic acid   elaidic acid

A fatty acid with more than one double bond, such as linoleic acid, can form a correspondingly larger number of geometric isomers, although only the *cis*/*trans* structures occur naturally.

trans/cis   trans/trans   cis/cis   cis/trans
                                        (natural form)

Possible isomeric forms of the linoleic acid chain

Other types of stereoisomerism also occur with unsaturated fatty acid structures. Position isomers such as linolenic acid and eleostearic acid have their double bonds located at different points along the hydrocarbon chain.

The chemical properties of the fats and fatty acids may be thought of in terms of the reactions of the carboxyl group and its attached hydrocarbon chain which can be either saturated or unsaturated. Although it is convenient to consider the reactions of these two parts of the molecule separately, it is important to remember that they exert a powerful influence on each other.

The carboxyl group of the fatty acids is involved in a number of reactions, perhaps the most important of these technologically being esterification and reduction. Some interesting synthetic fatty esters have been prepared for use mainly in cosmetics. Thus by treating long chain fatty acids such as myristic or palmitic acid with boiling methanol in the presence of a little concentrated hydrochloric acid, the methyl ester separates out as an oily layer. Iso-propyl myristate and iso-propyl palmitate are especially useful in reducing the viscosity and greasiness of other cosmetic fats. Similarly, synthetic glyceryl esters such as glyceryl monostearate have become widely used in the preparation of cosmetic emulsions.

Interesterification is often carried out in order to alter the characteristics of a fat. Heating to about 100 °C in the presence of a small amount of catalyst (0·2% sodium ethoxide) promotes interchange of the fatty acid radicals among the component glycerides. This random distribution affects properties such as melting point, grain structure and plasticity and is often used to improve the characteristics of edible fats.

Reduction of fatty acids using catalytic hydrogenation, or increasingly by the use of sodium in alcohol (Beauveault-Blanc reaction), is a valuable industrial route for the production of the synthetic long chain aliphatic alcohols required in the manufacture of synthetic detergents.

$$\text{/\\/\\}COOC_2H_5 \xrightarrow[\text{Na/EtOH}]{\text{hydrogenation}} \text{/\\/\\}CH_2OH + C_2H_5OH$$

fatty acid ethyl ester   primary   ethanol
                                   aliphatic alcohol

($\text{/\\/\\}$ represents a hydrocarbon chain)

Fats can be hydrolysed using water or steam to reform free fatty acids and glycerol (fat-splitting). If the hydrolysis is carried out using a caustic alkaline solution, water soluble soaps are formed. Fats are also decomposed at high temperatures (pyrolysis) to yield hydrocarbons. This process was used at one time in China for the production of motor fuel.

Saturated hydrocarbon chains are relatively inert, but the unsaturated chains are affected by both oxidizing and reducing agents. A certain amount of oxidation occurs when unsaturated fats and oils are exposed to the air (autoxidation). This is responsible for the 'drying' of oils such as linseed oil used in the manufacture of linoleum and paints, and for the rancid taste and smell which develop in fats. It has been shown that autoxidation involves attack by atmospheric oxygen at methylene groups adjacent to the double bonds.

$$CH_2CH{=}CHCH_2 \xrightarrow[\text{oxidation}]{\text{atmospheric}}$$

unsaturated region of fatty ester chain

$$CHCH{=}CHCH_2$$
$$OOH$$

hydroperoxide derivative

(represents a hydrocarbon chain)

Oxidation of unsaturated fatty acids and fats with oxidizing agents such as potassium permanganate, hydrogen peroxide and the halogens is mainly of analytical rather than commercial interest.

Prior to the onset of rancidity certain flavour changes are detectable in most oils and fats. The bland pleasant taste produced during processing often reverts to the original undesirable flavour of the crude oil. This phenomenon is known as *reversion* and is especially noticeable in fish oils. It has been found that reversion is particularly likely in fats such as soya bean oil which are rich in linolenic acid. The development of reversion is also influenced by short-wave light, temperature, oxygen tension and traces of metals such as copper and chromium which catalyse the reaction. For this reason metal 'scavengers' such as EDTA (ethylenediamine tetra-acetic acid) are often used during fat processing.

The catalytic hydrogenation of unsaturated oils is of great industrial importance in the manufacture of solid edible fats. The catalyst which is commonly used industrially is finely divided nickel, which forms an intermediate compound with the fat and gaseous hydrogen. This technique was first used commercially in 1906 after successful experiments carried out by Norman four years earlier. It is interesting to note that selective hydrogenation occurs across double bonds separated by activated methylene groups. Thus a linolenate with two activated methylene groups is preferentially hydrogenated over a linoleate with only one. Both are hydrogenated before oils such as oleates which possess only a single double bond, or oils with double bonds separated by more than one methylene group, neither of which have such highly active structures.

In addition to fatty acid esters and small amounts of free fatty acids the commercial oils and fats invariably contain other lipid substances. Crude seed oils contain quite appreciable quantities of phospho-

(*a*) $CH_3CH_2CH{=}CHCH_2CH{=}CHCH_2CH{=}CH(CH_2)_7COOH$ linolenic acid

(*b*) $CH_3(CH_2)_4CH{=}CHCH_2CH{=}CH(CH_2)_7COOH$ linoleic acid

(*c*) $CH_3(CH_2)_7CH{=}CH(CH_2)_7COOH$ oleic acid

(activated methylene groups $-CH_2-$)

lipids such as lecithins and cephalins, and sterols such as cholesterol. These are mostly removed during purification.

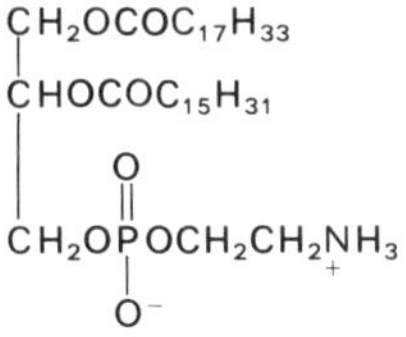

$CH_2OCOC_{17}H_{33}$
$CHOCOC_{15}H_{31}$
$CH_2OPOCH_2CH_2NH_3$

lecithin
(a phosphoglyceride)

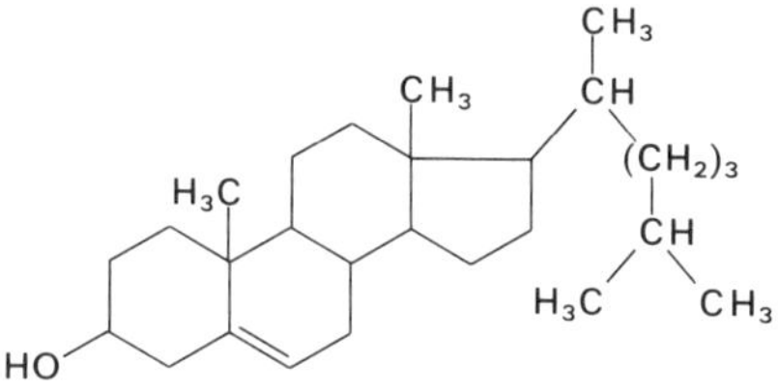

cholesterol (an animal sterol)

Occasionally unsaturated hydrocarbons such as squalene ($C_{30}H_{50}$) and pristane ($C_{18}H_{28}$) are also present. The darkening of oils during heating or storage is partly due to destruction of the tocopherols (vitamin E) which when present inhibit atmospheric oxidation. Thus

2

the stability of coconut oil at room temperature (5 mg tocopherol/ 100 g) is very much less than cottonseed oil (110 mg tocopherol/ 100 g). Other antioxidants are also thought to be present in natural fats since some relatively stable oils have a very low tocopherol content.

The formation of vegetable fats and oils is still shrouded in mystery. It appears that part of the carbohydrate material produced by photosynthesis is converted into saturated fatty acids which then react with glycerol to form saturated fats. It has been suggested that a further series of reactions then occurs producing unsaturated fats and oils and free fatty acids. Other evidence suggests that the saturated and unsaturated fats are formed quite independently.

Animal fat can originate from ingested carbohydrate, or be formed from dietary fat or protein. Experiments with rats and pigs have shown that certain 'essential fatty acids' such as linoleic and arachidonic acids can only be produced from ingested fat.

Animal and vegetable waxes also contain fatty acid esters, but these differ from those present in the fats in being formed from the higher monohydric alcohols such as cetyl alcohol ($C_{16}H_{33}OH$) and myricyl alcohol ($C_{30}H_{61}OH$) instead of glycerol. The vegetable waxes occur as deposits on the surface of certain plant leaves, stems and fruits, protecting them from excessive water loss. Thus candelilla wax is obtained from the stems of a species of Mexican weed, and carnauba wax, which is the hardest of all the commercial waxes, is recovered from the leaves of the Brazilian palm.

The majority of the animal waxes are obtained from marine sources, especially from the sperm and bottlenose whales. Beeswax and wool-fat are also commercially important. The purification of the crude waxes is usually carried out by melting in boiling water, and then bleaching the skimmed product.

ANALYSIS OF FATS

Physical Characteristics

Fats and oils can often be identified by their physical properties such as melting point, refractive index and density. Because of their complex nature and the existence of different crystalline forms the determination of melting point is not as precise as in the case of pure organic compounds. Most fats tend to supercool and soften over a range of temperatures and may even exhibit a double melting point because of conversion from one crystalline polymorph to another. The *slip point* is often measured by noting the temperature at which a plug of fat rises in a capillary tube or metal cylinder suspended in a heating bath.

Although there is generally an increase of specific gravity with chain length, differences are so small that refractive index is a more useful physical constant for identifying fats. Using a refractometer at constant temperature (25 ° or 40 °C is usual) the purity of samples can be rapidly estimated and their identification is possible. For cooking fats the temperature at which hazing occurs (*smoke point*), and vapour ignition point (*flash point*) are obviously of importance. If an oil is dissolved in a warm solvent in which it is sparingly soluble turbidity

		m.p. °C
Saturated fatty acids		
Myristic	$C_{13}H_{27}$ COOH	54
Palmitic	$C_{15}H_{31}$ COOH	63
Stearic	$C_{17}H_{35}$ COOH	70
Arachidic	$C_{19}H_{39}$ COOH	75
Benenic	$C_{21}H_{43}$ COOH	80
Lignoceric	$C_{23}H_{47}$ COOH	84
Unsaturated fatty acids		
Oleic $CH_3(CH_2)_7CH{=}CH(CH_2)_7$ COOH *cis*-9-octadecenoic acid		16
Linoleic $CH_3(CH_2)_4CH{=}CHCH_2CH{=}CH(CH_2)_7$ COOH *cis-cis*-9,12-octadecadienoic acid		−5
Linolenic $CH_3CH_2CH{=}CHCH_2CH{=}CHCH_2CH{=}CH(CH_2)_7$ COOH *cis-cis-cis*-9,12,15-octadecatrienoic acid		−11

Melting points of common fatty acids in most stable crystalline form

occurs on cooling (Crismer test). This *turbidity point* is of use in indicating adulteration of edible oils, especially as it is sensitive to free fatty acids.

	Smoke point °C	Flash point °C
Corn oil	227	326
Olive oil	199	321
Soya bean oil	256	326
Cottonseed oil	235	325

The viscosity of fats determined at a given temperature can be used to monitor chemical changes such as oxidation, degradation or polymerization which may occur during processing or storage.

The structure of the glycerides, fatty acids and fatty acid esters derived from fats has also been studied using conventional physical methods such as X-ray analysis, absorption and mass spectrometry and nuclear magnetic resonance (NMR). Thus X-ray diffraction patterns revealed the 'tuning fork' structure of triglycerides and together with spectrometric techniques enabled 'fingerprints' of individual glycerides to be built up. NMR has been used to detect isolated or multiple double bonds in unsaturated fatty acid chains.

Chemical Tests

Apart from the determination of physical properties and systematic chemical analysis, a number of chemical tests are carried out on fats to determine those characteristics which are of commercial importance.

The *saponification number* is the number of milligrammes of potassium hydroxide required to saponify one gramme of fat, and is usually in the region of 180 to 200. This figure is useful in soap making and for detecting adulteration.

The *acid value* of a fat is defined as the number of millilitres of 0·1 M potassium hydroxide required to neutralize a given (5 g) sample. The water soluble acids (butyric—C_4 to capric—C_{10}) are determined separately (*Reichert Meissl number*) from the water insoluble acids (*Polenske number*).

The *iodine number* is a measure of the degree of unsaturation of a fat and is defined as the number of grammes of iodine absorbed by 100 g of fat. Absorption is speeded up by using a catalyst such as bromine or chlorine. A convenient technique due to Wijs involves treating a solution of the fat in carbon tetrachloride with a known excess of iodine monochloride. After standing in darkness for about an hour the solution is back titrated with sodium thiosulphate to determine the amount of iodine absorbed. Highly unsaturated glycerides such as linseed oil and tung oil, with an iodine number from 180 to 190, are known as drying oils because they thicken in contact with air and finally form a tough elastic skin. Glycerides with a lower iodine number in the range 100 to 120, such as cottonseed oil, are called semi-drying oils and these thicken in air but do not form a skin. Non-drying oils such as olive oil (iodine number 85) and solid fats have iodine numbers which are often well below 100.

A number of tests are used to detect specific oils. Usually these depend upon the presence of small quantities of materials which do not form soaps with alkalis. These are often hydrocarbons or higher alcohols and are grouped together as unsaponifiable matter. As an example the Baudoin test involves refluxing the fat with concentrated hydrochloric acid and then adding a few drops of furfural in ethanol. A bright red coloration indicates the presence of sesame oil.

The presence of unsaponifiable matter can also be used to indicate the origin of a fat. The fat is saponified by boiling with potassium hydroxide solution, washed with water to remove soaps and the residue extracted with ether. On recrystallization the unsaponifiable matter yields crystals of a characteristic shape. Animal fats yield thin rhombic wafers of cholesterol, and plant fats, tufts of long needle-like crystals of phytosterol, a mixture of the two plant sterols, sito-sterol and stigmasterol.

Detection of rancidity usually involves testing for the presence of fatty acid breakdown products such as aldehydes, ketones and hydroperoxides. In the Kreis test a small quantity of phloroglucinol in ether is added to a little of the fat acidified with concentrated hydrochloric acid. A reddish coloration due to the presence of epihydrin aldehyde indicates rancidity.

The volatile carbonyls present in thermally abused fats can be removed by steam distillation. The distillate is passed through a solution of 2,4-dinitrophenylhydrazine which removes the carbonyl components as the corresponding phenylhydrazones.

Chromatographic Techniques

Chromatography is being increasingly used in the analysis of fats. Originally mixtures of fatty acids produced by hydrolysis of fats were separated and identified by their selective adsorption on a column of aluminium oxide. Identification was established by eluting the column with an indicator solution to reveal the acid layers or by using an ultraviolet lamp to make the separated components fluoresce.

More recently thin layer chromatography (TLC) and gas–liquid chromatography (GLC) have been successfully used to resolve mixtures of glycerides, fatty acids or their methyl esters. By these techniques the component glycerides of fats can be identified and estimated quantitatively. In TLC thin layers of silica gel or kieselguhr (325 μm) are spread on glass plates and spots of glyceride mixtures together with known 'marker' compounds are eluted with solvent mixtures. The addition of silver nitrate to the adsorbent slurry before spreading produces a better separation. This is due to the formation of weak coordination complexes between the silver ions and un-saturated ethylenic bonds present in the fatty acid components of the glycerides. To locate the separated glycerides the developed plates are sprayed with chromic or sulphuric acid and heated, or sprayed with dibromo-R-fluorescein and viewed under ultraviolet light. Unsat-urated components can also be revealed by exposing the developed plates to iodine vapour.

Separation or identification by GLC is preceded by conversion of the glycerides to their methyl esters by refluxing with a hydrochloric acid/methanol solution or treating with diazomethane in ether. The resulting esters are then volatilized by warming and passed through a heated 'Celite' column treated with a low volatile solvent such as polyethylene glycerol adipate. Elution is effected by a stream of inert gas and the components of the mixture recorded automatically on a chart as they reach the detecting device.

The separation of fatty methyl esters can also be carried out by repeated extraction with a mixture of two immiscible solvents (countercurrent distribution).

Enzymatic Hydrolysis

The enzyme pancreatic, lipase, promotes the hydrolysis of triglycerides, breaking them down ultimately to glycerol and free fatty acids. Unlike alkaline hydrolysis the 'outer' 1,3-radicals of the glyceride are attacked preferentially. If hydrolysis is stopped at about 65% completion the 2-monoglyceride can be isolated by solvent extraction. Alkaline hydrolysis followed by methylation can then be used to produce the corresponding methyl esters for chromatographic analysis. This type of enzymic hydrolysis is of great value in determining the chemical structure of fats.

THE FUTURE

The demand for fats and waxes continues to rise with rising standards of living and a world population which is increasing at the rate of 70 000 a day. A shortage of edible oils and fats seems to be especially likely in the developing countries of Asia and Africa within the next few years. Asia alone is likely to need about 70% more edible fats by 1973 than were required in 1968. In western Europe the position is less critical because of the increasing use of petroleum products for detergent manufacture and the much slower rise in consumption, estimated at 15% during the same period.

To cope with this position efforts are continually being made to increase the production of vegetable and animal fats. Improvements in agricultural methods, use of pesticides, fertilizers and new seed varieties have all played their part. Yields have been raised by improving extraction techniques. Chemical research has made possible developments such as hydrogenation and interesterification and has enabled more economical usage of the by-products of refining processes.

In the following sections some of the applications of fats and waxes are described in detail. The mineral oils and waxes and the silicones are omitted as they are dealt with elsewhere.

Chapter 2

Fat Technology

Primitive man used fat as a food and probably as a lubricant and crude skin medicament. Later, soap was discovered, and this was followed by the first attempts at cosmetics production. Fatty animal tissues were first rendered by cooking processes such as boiling and roasting. Vegetable oils were probably extracted by exposing heaps of oil-bearing material to the heat of the sun and collecting the exuding liquid. This crude form of extraction is still practised by primitive tribes today.

The ancient civilizations of Egypt, India and China are known to have used presses to extract vegetable oils. The plant material was first crushed in a mortar and then packed into baskets which were compressed by means of a wooden lever. The Romans and Greeks improved on this technique by introducing screw and winch presses. Wet rendering of animal fat was also well established by Roman times. A major advance was the discovery that oil yields could be improved by partly cooking the seeds before crushing and pressing. The heating of plant material to facilitate oil extraction is mentioned by Pliny at about this time. Speaking of the production of castor oil from the seeds he says: 'They must be boiled in water, and the Oil that swims on the top must be taken off. It is ill to eat, but good for Candles.'

Another description of the technique is given by the Greek physician Dioscorides, as described by Porta in 1658: 'Let ripe seeds, as many as you please, wither in the hot sun, and be laid upon hurdles: let them be so long in the Sun, till the outward shall break and fall off. Take the flesh of them and bruise it in a mortar diligently, then put it into a Cauldron glazed with Tin that is full of water: put fire under and boil them, and when they have yielded their inbred juyce, take the vessel from the fire, and with a shell skim off the Oyl on the top, and keep it. But in Egypt where the custom of it is more common: they clean the seeds and put them into a mill, and being well ground, they press them in a press through a basket.'

Crude pestle and mortar presses were also used for many centuries in the Far and Middle East for crushing oil seeds. The heavily weighted stone pestle was turned by an animal. In Europe 'stamper' presses were used until comparatively recently—especially in Holland. These contained the oil seeds in a filter cloth bag which was subjected to pressure by driving in wedges.

There was little change in these methods until the seventeenth and eighteenth centuries, when several improvements in the design of oil extracting machinery were made, notably the modern roller press invented by Smeaton, the builder of the first Eddystone lighthouse, in 1752. This was followed in 1795 by the invention of the hydraulic press by Joseph Bramah, a Yorkshire cabinet maker.

The later closed presses consisted of a piston moving in a cylinder which was slotted to drain off the oil. This dispensed with the need for the press cloth or fine basket originally used to hold the seeds during pressing.

More recently continuous screw presses, working on the principle of the domestic meat mincer, and continuous solvent processes have been introduced for the extraction of vegetable oils.

The first attempt to extract fats and oils by solvents was made by Jesse Fisher in 1843, and a solvent extraction process was patented by Deiss in 1856. It was not until later, however, that this process became commercially feasible, although by the outbreak of World War II very efficient continuous extraction plant was being widely used in Germany and the USA. Rendering is still the most widely used method of recovering animal and vegetable fats and oils.

Oil Milling

The most important sources of vegetable oils are seeds such as soya beans, cottonseed, groundnuts and linseed, together with the flesh of certain fruits such as the olive, palm and coconut. As all these have to be imported, the oil mills in the UK are situated near waterways or docks. Installations at Hull originally catered for shipments of oil seed from the Baltic, and imports from America, Africa and other areas were handled at London and Liverpool. After off-loading from the ships the oil-bearing seeds and fruits are stored in silos, the largest having a capacity of about 1000 tonnes and a height of 24 metres (80 ft). The daily throughput of a large oil mill can exceed

	Area of cultivation	World production (kilotonnes)	Oil content %	Uses
(a) Seed oils Soya	Far East, USA, Brazil	39 000	13–20	margarine, cooking fat, salad oil—meal used for pigs, cattle, poultry
Groundnut	Far East, W. Africa, Argentine	16 240	42–46	margarine, cooking fat, ice-cream—nuts eaten raw, cooked or as peanut butter—meal as cattle food
Cottonseed	USSR, USA, S. America, Africa, China	20 150	15–25	margarine, cooking fat, salad dressing—meal as fodder, fertilizer, fuel
Sunflower	USSR, France, Argentine, E. Europe	9 030	20–40	margarine, cooking fat, soap, paint, lubricators—meal for cattle, poultry—stalks, leaves for fodder, fuel
Rapeseed	India, China, Canada, France, Poland	4 999	35–40	table and cooking oil—cake for animal fodder
Sesame (Benniseed)	India, China, Mexico, Sudan	1 570	50	salad and cooking fat, preserving

Source	Location	Production	Oil %	Usage
Linseed	N. America, Argentine, USSR, India	3 115	30–40	paint, varnish, linoleum, oilcloth, printing ink—cake for animal feeding
Castor seed	Brazil, USSR, India, China, Thailand	731	42–46	paints, plastics, toothpaste, soap, varnish, laxative—cake for cattle feed after treatment
Maize	USA, Africa	300	3	table oil, cooking fat
(b) Nut oils Coconut	Philippines, Ceylon, India, Malaysia, Mexico	3 680	63 (copra)	soaps, lubricants, hair dressings, margarine, cooking fat, fuel—flesh desiccated for sweets, cooking
Oil palm	Nigeria	1 145 (palm-oil)	30–40	cooking fat, margarine
	Congo, Malaysia, Indonesia	950 (palm kernel oil)	50	cooking fat, margarine, sugar confectionery, soap
Tung	Argentine, China, USA	336	50	varnish, paints, enamel, plastics, lacquers, wood dressing
(c) Other sources Olive	Italy, Spain, Greece, N. Africa	1 344	12–15	table, cooking oil, soap—eaten raw or pickled

Usage and relative importance of principal vegetable oils

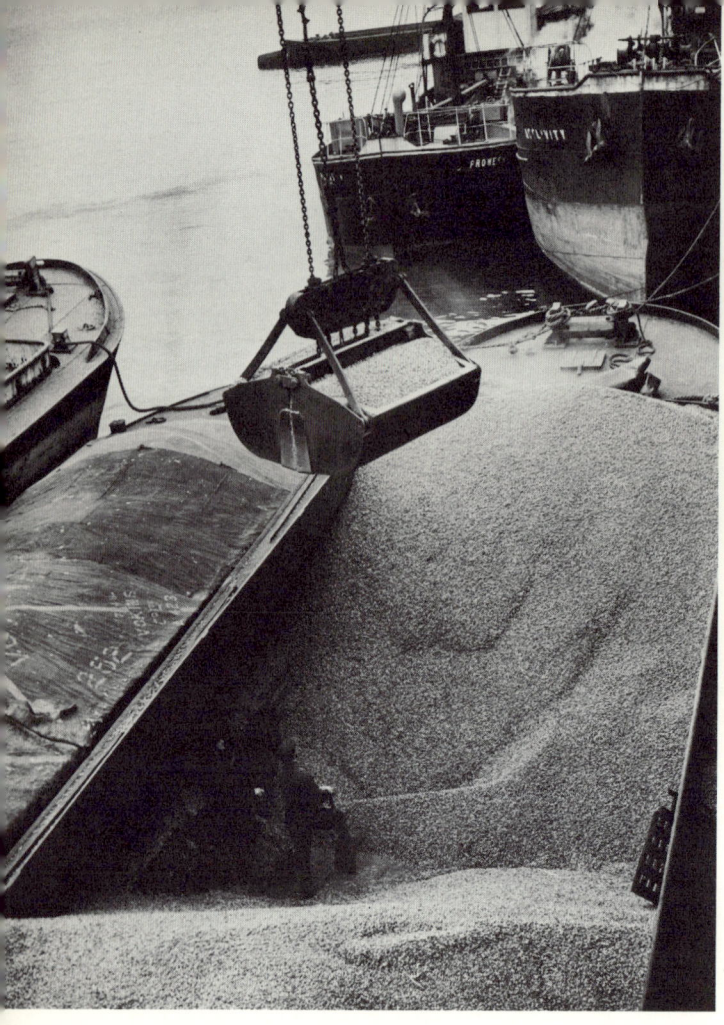

Off-loading groundnuts from a barge

2000 tonnes of oil together with an equal tonnage of oil cake for animal feeding.

Before pressing, the oil-bearing material is screened in a rotary sieve to extract dirt, leaves and other foreign matter, particles of iron being removed by a magnet. Additional preparation is often necessary in the case of seeds which may need to have seed hairs removed (e.g. delinting of cottonseed), or shells cracked off (e.g. decorticating of groundnuts). Tough skinned seeds such as palm kernels are passed through fluted break rollers and cereal seeds such as maize usually have the germ removed by passing between pairs of concentric

Stack cooker and screw press unit installed in a Greek oil mill

Installation of 12 screw presses for the extraction of edible oil from 12 200 kg
of oil seed per day per machine

tapered cylinders called 'degermers'. The screened and prepared material is then crushed to form a meal by passing between steel 'reducing' rollers.

In order to obtain the maximum possible yield the meal is then cooked at 70–110 °C in a steam-jacketed vessel to rupture the oil cells and adjust the moisture content. In the older hydraulic process the cooked meal was enclosed in filter-cloth packs and pressed to form cakes. The cakes were then transferred to a box press where the oil was expressed and drained into a settling tank prior to filtration.

This process has now been superseded by the continuous screw press or expeller in which the meal is forced through a tapered horizontal tube by means of a revolving Archimedean screw. The oil is forced through fine slits varying in width from 0·25 mm ($\frac{1}{100}$ in) to 1·25 mm ($\frac{1}{20}$ in) located along the body of the press. The pressed oil cake which is ejected from the end of the tube contains about 20% of the original oil content. Low pressure expellers of this type have a capacity of about four tonnes of seed an hour and the oil produced is of high quality. High pressure expellers which remove up to 95% of the oil have a much smaller throughput (0·5–0·75 tonnes/hour) and produce a lower quality oil containing cell debris and other material. For this reason extraction of the meal from low pressure expellers is more usually effected by solvent extraction which can also be used for the whole process.

Two techniques are in use. In the older percolation process the prepared meal is charged into perforated troughs which are carried up and down through the hot solvent by means of a travelling band to which they are attached. The later Hildebrandt or 'total immersion' process conveys the meal against the flow of solvent ('counter-current' extraction) by means of a screw thread perforated with small holes. The most common solvents are trichloroethylene and n-hexane, although benzene and other organic solvents have been used. The solvent is recovered from the oil solution and the extracted meal by means of a small distillation unit ('stripper'). Most of the exhausted oil cake is used for cattle feed, with the exception of a few materials such as tung oil meal and castor oil meal which are poisonous to cattle and are therefore used as fertilizers.

A variation on the countercurrent process is commonly used in the UK in which the rolled ('flaked') seed is carried along a moving belt

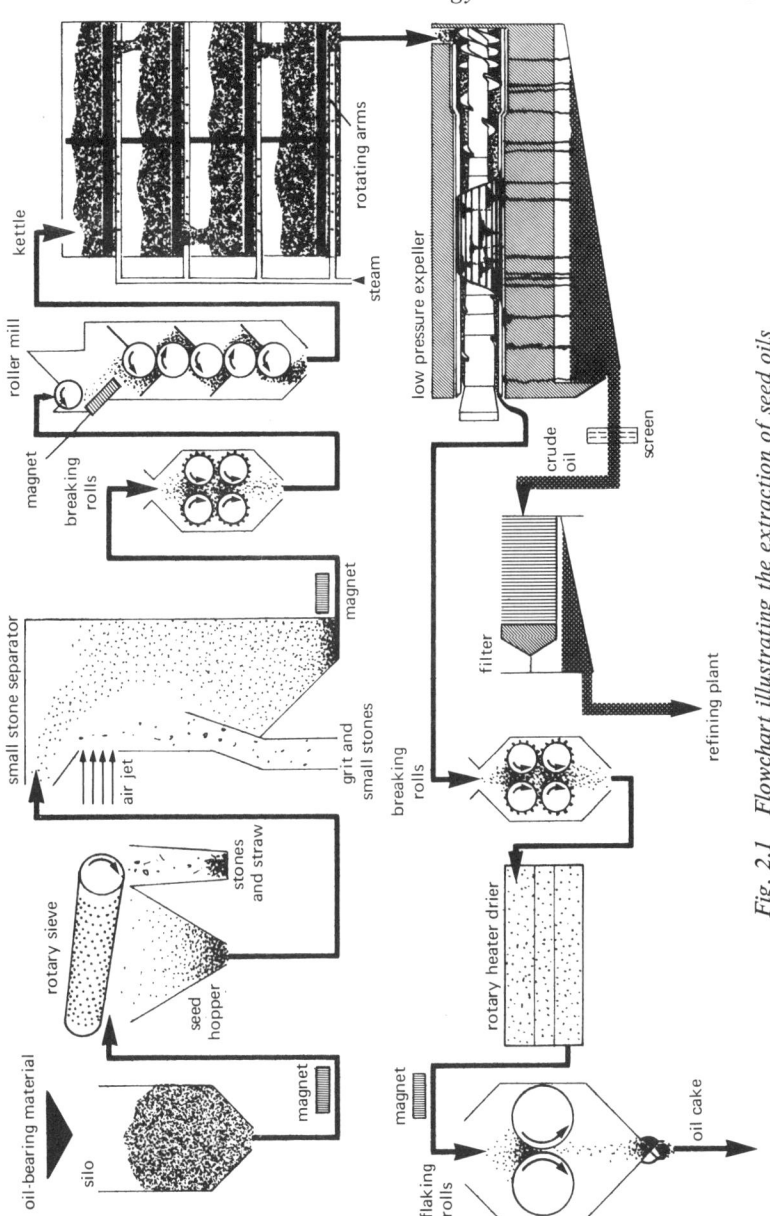

Fig. 2.1 Flowchart illustrating the extraction of seed oils

of fine wire mesh beneath a succession of solvent sprays. The oil-bearing solvent, usually petroleum ether, is passed from spray to spray in the opposite direction to the moving belt. In this way fresh solvent washes the almost completely extracted cake at the end of the belt. The final solvent spray washing the flaked seed at the inlet is heavily loaded with oil. It is pumped to the distillation plant where the petroleum ether is recovered and the crude oil is passed to the refining stage.

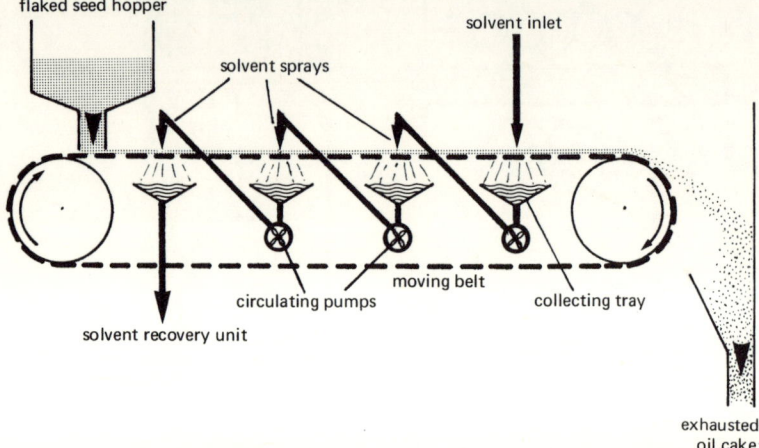

Fig. 2.2   Countercurrent solvent extraction of flaked oil seed

Extraction of Animal Fats

The recovery of animal fats is carried out by three types of rendering process. The oldest method of 'open kettle' rendering involves heating in open pans of water over a fire. This has now been replaced by wet or dry rendering.

In *wet rendering* the selected, finely chopped material is charged into vertical cylinders with conical bases to assist in drawing off the water after settling. Steam is then blown through the cylinders for several hours, the pressure being kept constant at about $3 \cdot 5$–5 kg/cm^2 (50–70 lb/in^2) by venting excess steam through a pressure valve.

After allowing settlement to take place, the floating fat is drawn off leaving the water ('stick water') and exhausted animal matter ('tankage') behind. This process is widely used for the extraction of edible fats such as lard, the stick water being evaporated to form an animal food additive. If the oil content of the tankage is too high it cannot be used for cattle food and is turned into fertilizer, unless specifically treated by a solvent extraction process to reduce the oil content to about 1%.

Whale oil from the blubber, meat and bones of the baleen whale is often extracted aboard specially fitted out factory ships, the chopped material being steam heated in giant pressure cookers. Fish oil from the herring and the related menhaden is obtained by screw pressing after steam cooking. The liver oils from the cod, halibut, shark and whale are extracted aboard the fishing vessels by steaming the chopped raw liver in steel boilers and running off the floating oil layer.

Dry rendering is carried out in large steam heated tanks under vacuum. During the heating process the chopped animal tissue is churned by metal blades (agitators). The fat cells rupture, allowing the molten fat to drain off from the bottom of the tanks. Any fat remaining in the residue is removed by pressing or solvent extraction. Although dry rendering is cheaper than steam rendering, the product is somewhat inferior and the process is restricted to the extraction of non-edible fats.

Fat Refining

The crude fats produced by the above techniques contain impurities which if not removed would give the product an undesirable taste, smell or appearance. These include free fatty acids, gummy residues, lecithin, protein and carbohydrate material, together with various coloured and odiferous substances. The extent of the refining process depends upon the product. In the case of high quality fats such as lard, all that is required is filtration to remove suspended particles after coagulation by steam. Other products such as palm oil also need to be neutralized, bleached, deodorized and winterized—processes which are described below.

Neutralizing or deacidification is commonly carried out by removing the free fatty acids as soaps by treatment with sodium

A 'Rotocel' solvent extraction plant for processing a variety of vegetable oil seeds

A solvent extraction plant in Holland for the removal of fat from 150 tonnes per day of meat, bone greaves, offal, etc.

hydroxide solution, a process known as *alkali refining*. The fat and alkali are agitated at about 90 °C for a short time until the fatty acids have been saponified. After standing for some hours, the soap settles to the bottom, carrying down suspended impurities including much of the colouring matter. This soapy layer ('foots') is run off, and the purified fat washed with hot water and then vacuum dried. Remaining traces of colour and smell are removed by treating the molten fat with about 1% of Fuller's earth which is filtered off after use.

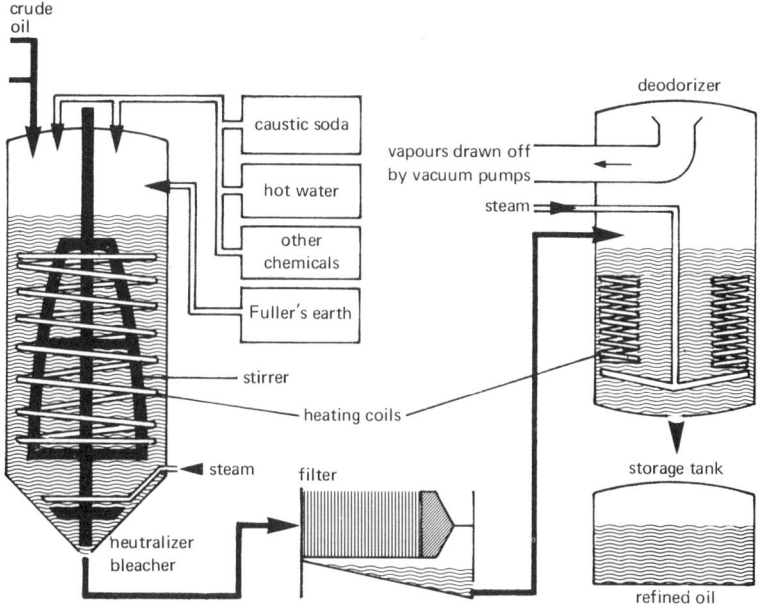

Fig. 2.3 Flowchart illustrating the refining of crude seed oils

Alkali refining has the drawback of being a batch process and *continuous deacidification* processes have been in use for some time. A gum conditioner is added to the liquid fat to reduce its solubility in the gummy impurities present; the fat is then continuously stirred into a carefully controlled proportion of sodium hydroxide solution in a mixing unit. This is maintained at a temperature of 20 °–32 °C. The mixed liquids emulsify and pass out of the mixer within about one minute. The emulsion is then broken by raising the temperature to about 70 °C and spinning in a high speed centrifuge to separate the soapy layer from the fat. The neutral oil is washed and centrifuged twice more before vacuum drying. This is sometimes called the 'Low Loss' process, since losses of fat during refining are about 20% less than with the batch method.

Continuous deodorizing can be carried out by passing steam up stainless steel towers fitted with trays and bubble caps (see *Fuels, Explosives and Dyestuffs*, Book 3 in this series), hot oil cascading down the towers coming into intimate contact with the ascending steam (countercurrent flow). Alternatively the fat is heated under vacuum for several hours using superheated steam. Palm oil and certain other vegetable oils can be bleached in a similar way using a hot air countercurrent technique.

Winterizing involves the removal from the oils of high melting point fractions which in cold weather would precipitate out and spoil the appearance of the product. The process is carried out by chilling the oil and then passing it through a filter press to remove any solidified fats.

Hardening of Fats

In 1896 two French chemists, Sabatier and Senderens, discovered that certain unsaturated organic substances which were normally difficult to hydrogenate, could be readily reduced in the vapour state by mixing with hydrogen gas and passing over freshly reduced nickel powder (vapour phase catalytic hydrogenation). Norman patented a process for hydrogenating oils and low melting point fats by this method in 1902. The British patent rights for hardening fats were bought by the Crossfield Co. in 1905, who then sub-licensed rights to several companies in other countries such as Jurgens in Holland, and Procter and Gamble in the USA.

This discovery was of great importance not only because oils and soft fats could be converted into more valuable hard fats (hence the term 'hardening'), but also because the process deodorized oils such as fish oils which had strong smells. Another advantage was that the

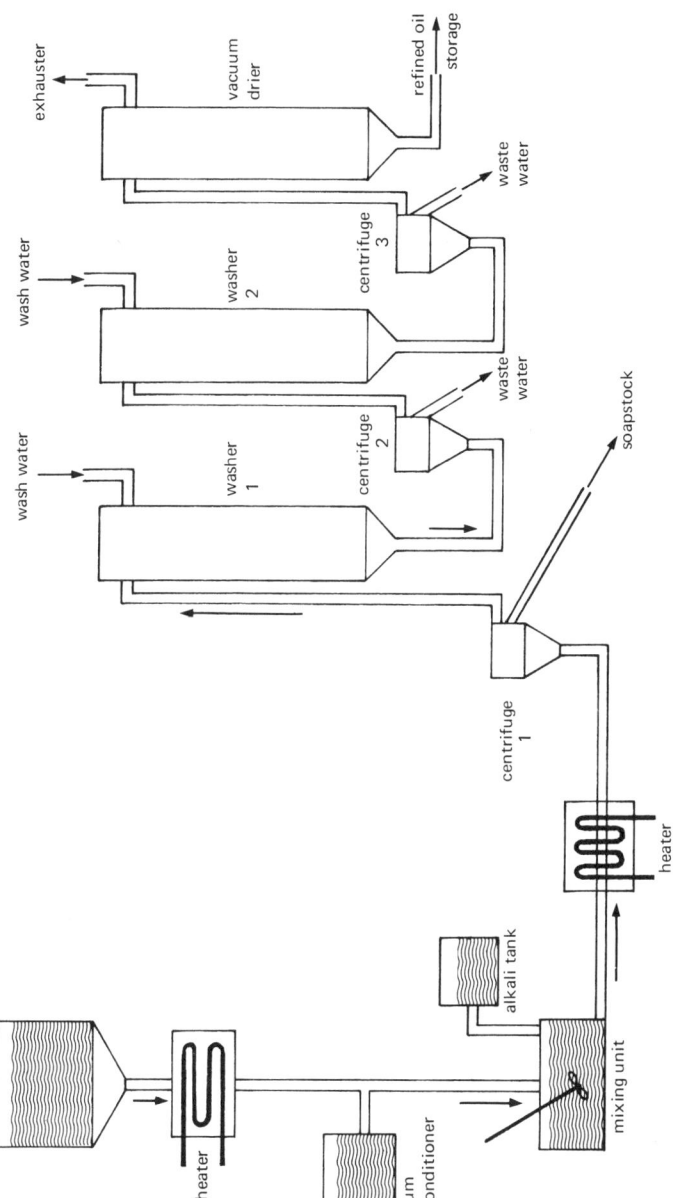

Fig. 2.4   Flowchart of continuous oil refining plant

increase in saturation of the hardened fats lessened their tendency to become rancid, the unsaturated double bonds permitting attack by atmospheric oxygen to produce the acids and aldehydes associated with the smell and taste of rancid fat.

The nickel catalyst must be prepared so as to provide as many 'active centres' as possible at which catalysis can occur. Impurities such as lead, arsenic and sulphur exert a poisoning effect on the catalyst. It is prepared in a pure, finely divided state by reducing the oxide at 300 °C in an atmosphere of hydrogen, or by precipitating nickel carbonate on the surface of a carrier such as diatomaceous

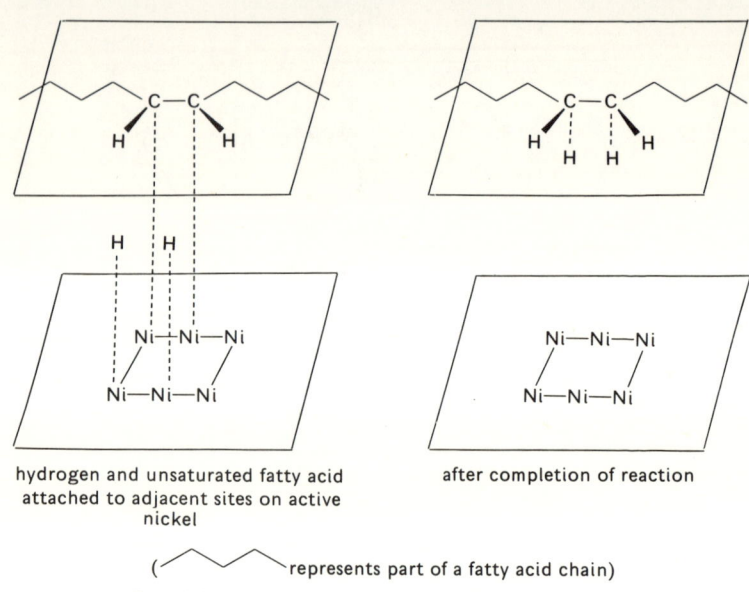

hydrogen and unsaturated fatty acid attached to adjacent sites on active nickel

after completion of reaction

($\diagdown\diagup\diagdown\diagup\diagdown$ represents part of a fatty acid chain)

Probable mechanism of nickel catalysed hydrogenation

earth and then reducing with hydrogen as before. In wet reduction processes a nickel salt such as nickel formate is intimately mixed with the unsaturated oil. On heating the mixture and passing hydrogen through it metallic nickel is produced in a finely divided state and hardening takes place as before.

Although continuous hydrogenation processes are in use, batch processes are still of major importance. Before hardening, the oil is pretreated by alkali refining and bleaching to remove possible catalyst poisons. The hydrogenation towers or converters are tall cylindrical steel pressure vessels with cone-shaped bases, fitted with stirring devices and heating and cooling coils. The mixture of pretreated oil and nickel catalyst is pumped into the converter, where it is partially mixed by stirring and partially by the flow of hydrogen entering at the base through perforated pipes. Steam is passed through the heating coils, while the oil mix is continually pumped to the top of the converter from where it is sprayed back down the tower. As the reaction is exothermic, the steam heating is stopped as the reaction gets under way, and the temperature is kept constant at about 200 °C by passing cooling water through the coils if required. The reaction most readily takes place with a high catalyst concentration and moderately low pressure of about 280–525 kN/m^2 (40–75 lb/in^2).

Hydrogenation of the fat charge is followed by determination of the melting point, refractive index, or iodine number and when hardening has taken place to the required degree, the reaction is

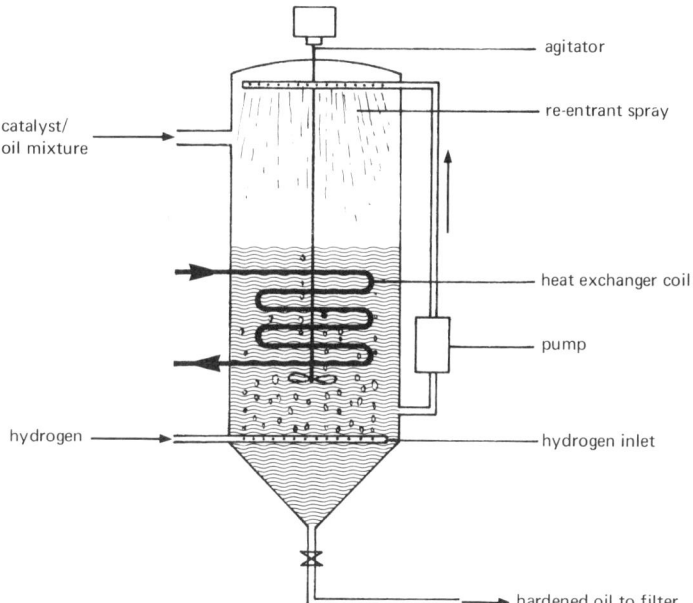

Fig. 2.5   Fat hardening converter tower

stopped by lowering the temperature to about 70 °C and filtering the catalyst. The product is rebleached, and deodorized under vacuum if required for human consumption. Whale oil, groundnut oil and soya bean oil are commonly hardened in this way for the manufacture of margarine and cooking fat.

At one time most of the hydrogen used for fat hardening was produced by passing steam over iron, which had been specially prepared by reducing iron oxide in a stream of water gas. More recently increasing use has been made of refinery gas containing short chain hydrocarbons. This is treated with steam in the presence of a catalyst to produce a mixture of hydrogen and carbon monoxide.

$$C_2H_6 + 2H_2O \xrightarrow{\text{catalyst}} 2CO\uparrow + 5H_2\uparrow$$
$$\text{ethane} \quad \text{steam}$$

EDIBLE FATS

The majority of animal and vegetable fats produced are used in the manufacture of 'edible fats'. Margarine has become established in the national diet and 400 000 tonnes are consumed annually in the British Isles, in addition to large quantities of cooking fats and shortenings. Before outlining the manufacture of edible fats it is interesting to examine their structure as this has an important bearing on their physical properties.

The edible fats are described as plastic solids although the percentage of solid fat present is often quite small. This reflects the ease with which they can be moulded by pressure, a property which is due to their structure. The solid glycerides present in edible fats are in the form of interlocked minute crystals, which hold trapped between them minute colloidal droplets of oil, water or other liquids. In addition, up to 15% of air is occluded in the fatty network. The plasticity of the product is varied by altering the composition, volume and crystal size of the solid glyceride content. Thus for confectionery, a fat with a fairly precise melting point and one which is plastic over a relatively short range is preferred. For margarine and cooking fat a product is required with a wide plastic range and indeterminate melting point. This facilitates spreading on bread and ensures an even distribution throughout cooking doughs. Variation of the glyceride content can also be made to suit the fat for different climatic conditions. During the early 1960's the USA produced what was called a 'Global Spread' margarine for the armed forces, which, it was claimed, had easy spreading properties over a very wide range of temperatures. 'Spread from the fridge' margarines of this type are now becoming popular in the UK.

To obtain the desired characteristics in the final product the conditions under which the hardening of oils takes place must be rigidly controlled. Thus hydrogenation of oils destined for the manufacture of margarine is carried out at low pressure (35 kN/m^2 (5 lb/in^2)), comparatively high temperature (about 200 °C), and with a selective catalyst aimed at maximum conversion of linoleates to oleates and minimum production of stearates. Cooking fat on the other hand is prepared by hydrogenation of suitable oils at moderately high pressure (420 kN/m^2 or 60 lb/in^2) at a temperature between 90 ° and 150 °C, and using a catalyst designed to suppress the formation of oleates. Recently the hydrogenation of edible fats has again been the centre of interest since medical research has shown that the inclusion of unsaturated fats in the diet reduces the incidence of cardiovascular disorders in humans. As a result intensive work has been carried out on the use of hydrogenation catalysts other than nickel, in order to preserve polyunsaturated components such as linoleic acid.

BUTTER

Butter is commonly churned from cream which has been ripened, although small amounts of 'sweet' butter are still made from fresh pasteurized cream. The ripening process is initiated by adding a small quantity of a starter culture to the cream after adjusting the pH with sodium bicarbonate and then pasteurizing. After about 14 hours at 150 °C the lactic acid content rises to about 0·4% and a number of flavouring compounds are produced by bacterial action. These include diacetyl which is synthesized from small amounts of citric acid in the cream by *Streptococcus citrovorus*. The effect of churning is to mechanically agitate the cream so that the oil-in-water emulsion becomes inverted to form a water-in-oil emulsion and granules of butter separate from the watery butter-milk. After the butter-milk has been drained off the butter is washed and from 2 to 10% salt is worked in. This enhances the flavour and reduces the possibility of bacterial infection. Continuous butter-making processes have been introduced during the last few years.

Unlike margarine the consistency of butter and its nutrient content are affected by seasonal variations in feeding stuffs and other factors associated with the cow such as breed, age and health.

MARGARINE

The Industrial Revolution which occurred in Europe during the nineteenth century brought with it a population explosion and a drain of workers from the land. This caused a fat shortage which was serious enough for the scientifically inclined Emperor of France, Napoleon III, to offer a prize for the inventor of a butter substitute. As a result a French chemist, Mège Mouriés, produced a synthetic butter which was patented in 1869. After studying the cows on one of the Imperial farms at Vincennes, Mouriés decided that the cream in milk was produced from the body fat of the animal mixed in some way with milk and water. By mincing beef tallow with warm salty water containing pepsin, and allowing it to stand for several hours, he was able to separate off a yellowish crystalline fat from which could be pressed out a soft semi-liquid fraction, later called oleo. The oleo was warmed and churned with milk to which had been added an extract of dried cow's udder—a special ingredient which was to convert the mixture into butter! Later the use of mammary gland extract was abandoned and sour milk was used instead. Also the use of expensive pepsin was dropped in favour of other rendering techniques. The emulsified fat produced by Mège Mouriés presented on cooling a characteristic pearly appearance from which the name margarine is obtained (Gk. *margarites*— pearly). The first factory for producing the new product was built at Poissy on the outskirts of Paris in 1869 but the venture was a failure. In 1871, however, two Dutch butter merchants Van den Bergh and Jurgens took up the process and within a decade margarine had become firmly established as a butter substitute. In Britain the new product was at first marketed under the name of 'Butterine' but this was altered to 'Margarine' by legislation in 1887 to avoid confusion with butter.

Steam rendering fats for margarine manufacture in the 1870's

Originally the churned emulsion was cooled by adding water and then kneaded to an acceptable consistency. Later, cooling was carried out by spraying chilled water on the emulsion as it was emptied from the churn and egg yolk was added as an emulsifying and colouring agent. In 1907 a new method of cooling was patented in the UK which involved spraying the margarine emulsion onto the surface of a large revolving metal drum, which was chilled internally with ice-cold brine. The solidified margarine was stripped from the drum in small flakes before being milled.

Modern margarine-making plant makes use of a device known as a votator, which was introduced in the USA in 1936. This is a highly efficient heat exchanger which cools the margarine emulsion out of contact with the air. It consists of a cooled, heat insulated stainless steel cylinder which contains a rapidly revolving solid shaft. This is fitted with scraper blades which press against the cylinder wall by centrifugal force and scrape off the chilled product which passes along the annular space around the shaft.

The blend of oils used for margarine is mainly vegetable in origin, about 80 000 tonnes each of groundnut oil, coconut oil, palm kernel

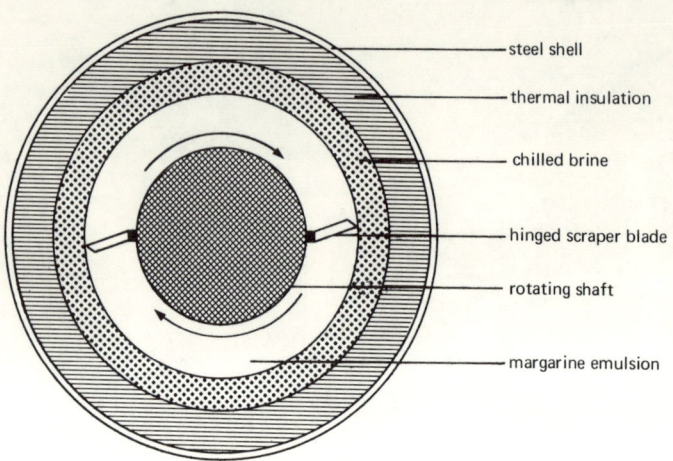

steel shell

thermal insulation

chilled brine

hinged scraper blade

rotating shaft

margarine emulsion

Fig. 2.6   Cross section of votator (scraped surface heat exchanger)

oil and palm oil being used annually in this country. In addition small amounts of whale oil and oleo oil are used. A suitable blend of refined soft fats and hydrogenated oils is chosen to give a product with the required degree of hardness.

The aqueous blend is prepared from pasteurized skimmed milk. This is 'conditioned' by adding a culture of lactic acid-forming bacteria, souring ('ripening') being allowed to continue until the required flavour has developed. Sometimes diacetyl ($CH_3 \cdot CO \cdot CO \cdot CH_3$) is added to give a butter-like flavour, together with a little salt and colouring agents such as carotene and annatto. Small quantities of lecithin or monoglycerides such as glyceryl monostearate are also often included and have the dual role of emulsifiers and anti-spattering agents, i.e. they prevent the fat 'spitting' when heated.

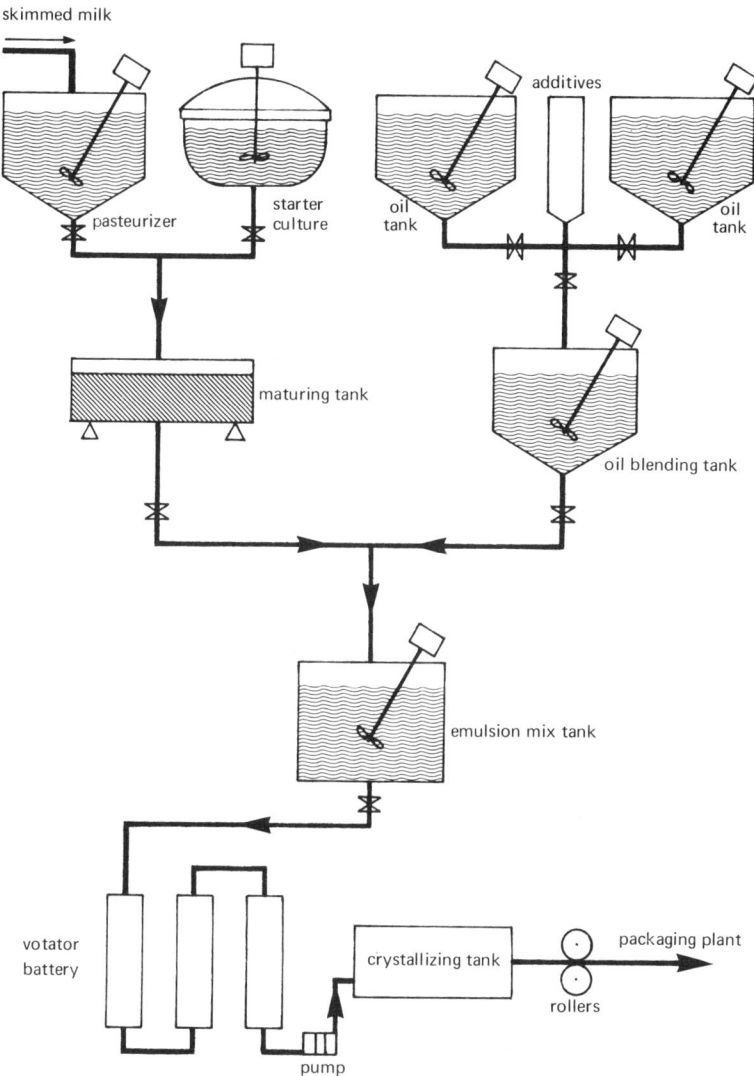

Fig. 2.7   Flowchart of margarine plant

Since 1927 vitamins A and D have been added to margarine blends, as these fat soluble vitamins are only found to a very limited extent in vegetable fats. The amounts added are controlled by law in the UK at 760–940 international units of vitamin A and 80–100 international units of vitamin D per ounce (28·3 g) of margarine. Preservatives are not allowed in Britain, nor must the water content of margarine exceed 16%.

An interesting feature of the flavouring of margarines is that the taste of the milk blend is only detectable if the fat has a melting point low enough to allow it to melt quickly in the mouth and hence release the occluded liquid phase of the emulsion. Attempts to imitate the flavour of butter have been quite successful.

The fatty and aqueous blends are vigorously churned to form a creamy emulsion using stainless steel cylinders fitted with counter-rotating paddles. After chilling the emulsion by passage through a battery of votators, the product is allowed to 'rest' for several hours. This maturing process makes it more suitable for working up into a smooth buttery texture by passing through a series of power driven rollers. At this stage butter may be added although the amount is limited to 10% by law.

COOKING FATS

The cooking fats differ from margarine and butter in being almost pure fat, and not fatty emulsions. Lard is produced by rendering pig-fat, the quality and composition varying with the method of production. Hydrolysis of lard yields both saturated and unsaturated fatty acids. Oleic acid forms almost half of the unsaturated acid content, the main saturated fatty acid present being palmitic acid. Attempts have been made to improve the plasticity of lard by re-shuffling the ester groupings of the glycerides present, a process known as randomizing or interesterification. This is effected by either heating with a catalyst such as sodium methoxide ($CH_3 \cdot O \cdot Na$) for a few minutes, or cooling the vacuum dried product to just above its melting point and then adding a liquid catalyst of sodium-potassium alloy. The semi-liquid lard is then pumped into crystallizing tanks where the higher melting point glycerides crystallize out at a temperature of about 37 °C. The catalyst is removed by treating the

Ripening tanks used in the manufacture of margarine

Margarine blending vats

lard with carbon dioxide and water, washing out the soap formed and finally centrifuging.

Lard is extensively used as a shortening agent, so-called because it imparts a short, crispy structure to cakes, shortbread and biscuits. Lard substitutes or cooking fats have now replaced lard to a considerable degree. There are several reasons for this, one being the inability of the lard industry to supply the enormous demand for shortenings. In addition lard substitutes can be prepared with a superior texture and resistance to rancidity, and a higher 'smoke point' for frying than the natural product.

Originally lard supplies were eked out by adulteration with animal stearins and cottonseed oil, and then lard substitutes were prepared which were composed solely of vegetable fats. Modern cooking fats are produced by blending about 80% of soft fats such as groundnut and palm oil with about 20% of hardened whale oil. This is then processed in a votator in a similar manner to margarine. After being allowed to crystallize, the white plastic solid is forced through an extrusion valve to improve its texture. It has been common practice for some years to add an emulsifying agent to cooking fats in quantities of up to about 10%. The material usually used is glyceryl monostearate, and fats so treated are termed superglycerinated. Superglycerinated fats are able to form emulsions with a high water ratio, and this in turn enables a higher proportion of sugar to be mixed with them. For this reason they are sometimes described as 'fat-extended' or 'high ratio'. The presence of emulsifiers such as glyceryl monostearate also improves the shortening characteristics of a cooking fat and its creaming properties (i.e. its capacity for incorporating air bubbles when beaten).

Glyceryl monostearate is also used instead of the traditional and more expensive emulsifying agents such as egg yolk in a number of prepared food products containing fat, such as mayonnaise and ice-cream. It is manufactured by heating a hardened fat, containing a high proportion of mixed triglycerides of stearic and palmitic acids, with glycerol in the presence of a catalyst such as sodium carbonate, or zinc or magnesium oxide. The mixture is heated with agitation at 180°–240 °C in a stainless steel or nickel reactor until reaction is complete, as indicated by the disappearance of the glycerol. It will be seen that the process involves 'sharing' out the available ester units to produce mono- and di-esters in place of the original glyceryl tri-esters. The glyceryl monostearate is finally purified by distillation at reduced pressure.

$$2 \begin{array}{c} CH_2OH \\ | \\ CHOH \\ | \\ CH_2OH \end{array} + \begin{array}{c} CH_2OCO(CH_2)_{16}CH_3 \\ | \\ CHOCO(CH_2)_{16}CH_3 \\ | \\ CH_2OCO(CH_2)_{16}CH_3 \end{array} \xrightarrow[+ \, Na_2CO_3]{180\,°-240\,°C} 3 \begin{array}{c} CH_2OCO(CH_2)_{16}CH_3 \\ | \\ CHOH \\ | \\ CH_2OH \end{array}$$

| glycerol | glyceryl tristearate (SSS) | glyceryl monostearate (GMS) |

The shelf life of lard and lard substitutes can be considerably increased by the use of antioxidants which delay the onset of rancidity. The most commonly used are butylated hydroxyanisole (BHA) and propyl, octyl and dodecyl gallates. The amount of antioxidant added is controlled by the 'Antioxidant in Food Regulations (1958)' and is usually in the region of 5–20 parts per million (ppm). There has been recent evidence that antioxidants in food may have a harm-

| propyl gallate (PG) | butylated hydroxyanisole (BHA) |

ful effect on young children by interference with tissue oxidation mechanisms. As a result the Food Standards Committee Report on Antioxidants in Food (1963) recommended that antioxidants should be prohibited in baby foods.

SURFACE COATING OILS

Certain unsaturated oils such as linseed oil and tung oil oxidize when exposed to the atmosphere in a thin layer, forming a tough but flexible solid film. Oils which behave in this fashion are termed drying oils, the drying power of the oil increasing with the degree of unsaturation. The mechanism of oil 'drying' is complex and not yet

4

fully understood, but is believed to involve the formation of hydro-peroxides by addition of oxygen to carbon atoms activated by an adjacent double bond.

(a) $\quad$ —CH=CH— $\xrightarrow[\text{atmospheric oxidation}]{+\ O_2}$ —$\overset{.}{C}$H—CH—
$\qquad\qquad\qquad\qquad\qquad\qquad\qquad\qquad\qquad$ |
$\qquad\qquad\qquad\qquad\qquad\qquad\qquad\qquad\qquad$ OO$^\cdot$

unsaturated
region of $\qquad\qquad\qquad\qquad\qquad\qquad\qquad$ double
fatty acid chain $\qquad\qquad\qquad\qquad\qquad\qquad$ free radical

(b) $\quad$ R$\overset{.}{O}$O + =CH$\overset{.}{C}$H$_2$CH= $\longrightarrow$ ROOH + =CH$\overset{.}{C}$HCH=

free $\qquad$ activated $\qquad\qquad$ hydroperoxide $\qquad$ free
radical $\quad$ methylene group $\qquad\qquad\qquad\qquad\qquad\qquad$ radical

$\qquad\qquad\qquad\qquad\qquad\qquad\qquad\qquad\qquad$ OO$^\cdot$
$\qquad\qquad\qquad\qquad\qquad\qquad\qquad\qquad\qquad$ |
(c) $\quad$ =CH$\overset{.}{C}$HCH= + O$_2$ $\longrightarrow$ =CHCHCH= $\longrightarrow$ polymerization
$\qquad\qquad\qquad\qquad\qquad\qquad\qquad\qquad\qquad\qquad\qquad\qquad\qquad\qquad$ or cleavage
free radical $\qquad\qquad\qquad\qquad\qquad\qquad$ free radical
$\qquad\qquad\qquad\qquad\qquad\qquad\qquad\qquad\qquad$ peroxide

Polymerization of the hydroperoxides takes place after a partial reshuffling of bonds along the carbon chains to give a pattern of alternate single bonds and double bonds (conjugation). Oils such as tung oil which already possess conjugated bond structures therefore polymerize very readily. Finally, cross linking of the polymer occurs, the remaining oil becoming trapped in the resulting network to give a tough gel. Oxidation of an oil film in this way begins slowly but rapidly accelerates until a large proportion of the unsaturated components have been accounted for, when the reaction rate falls off. Vegetable oils contain small quantities of natural antioxidants and it is the presence of these which initially slows the drying process.

The oxidation of oil films can be speeded up by the addition of chemical accelerators which are known as 'driers'. These are usually mixtures of lead, manganese or cobalt naphthenates dissolved in white spirit. Rosin salts and linoleates produced by the hydrolysis of linseed oil are also used as driers.

The most important of the drying oils is linseed oil, accounting for almost half the annual world consumption. The oil is produced from the seeds of the flax plant which is grown throughout the temperate zones of the world, annual production running at about 4 500 000 tonnes (10 thousand million pounds). The largest producer is the

USA, followed by the Soviet Union. The expressed oil is refined by treating with alkaline liquor (alkali refining) or stirring with 1% sulphuric acid. Increasing use is also being made of solvent extraction. The refined 'raw' oil is relatively slow drying and does not give a glossy film. Its properties can be greatly improved by suitable heat treatment. 'Boiling' is carried out by heating the raw linseed oil in large ventilated coppers (kettles) using superheated steam coils. Air is blown into the oil at intervals and at 100 °C mixtures of driers are added. The temperature is allowed to rise to between 125° and 150 °C, the contents then being rapidly cooled by passing cold water through the steam coils. The resulting 'boiled oil' has a greatly reduced drying time, is more viscous and produces a superior gloss film compared to the raw product. Blown oils are produced by prolonged air blowing during the boiling process. These have good drying properties but produce an inferior film, and are used for undercoats and flat paints.

An alternative to boiling is the stand oil treatment which involves heating the refined oil in the absence of air to a temperature of about 300 °C for several hours. Formerly the oil was heated in open kettles until the hot vapour ignited, this was then left to burn until the processing was complete. The product in this case was known as 'top-fired' oil, but is rarely used now except for certain types of printing inks. Stand oil is widely used for making gloss paint, varnish and printing ink.

Other drying oils, with the exception of conjugated oils such as tung oil, are also subjected to heat treatment, using the same techniques as for linseed oil. In addition, solvent extraction processes are occasionally used with oils such as soya bean oil and fish oils to isolate the more highly unsaturated fraction of the glycerides present. The solvent used is usually either liquid propane (Kellogg 'Solexol' process) or furfural—an aldehyde extracted from cereal husks and corn cobs.

furfuraldehyde (furfural)

Attempts have been made to improve the drying qualities of oils by promoting conjugation of the double bonds—a process known as 'isomerization'. This has been done on a small scale by prolonged heating with a nickel catalyst at temperatures around 160 ° to 170 °C. The cost is too high for widespread commercial application, however.

Tung oil, with its rapid drying properties, is especially useful in the manufacture of varnish and was originally second in importance to linseed oil. It is extracted from the seeds of a tree native to China, and difficulties of supply during World War II stimulated the production of synthetic drying oils, such as dehydrated castor oil, which have now superseded it in importance.

Castor oil is a non-drying oil which yields a high proportion of the hydroxy acid ricinoleic acid on hydrolysis. By heating ricinoleic acid at high temperature and low pressure in the presence of a dehydrating catalyst such as phosphoric acid, a conjugated double bond structure is produced giving rise to a drying oil known as dehydrated castor oil.

$$\text{OH}$$
$$\text{CH}_2\text{CHCH}_2\text{CH}=\text{CHCH}_2 \xrightarrow[\text{elevated temperature}]{\text{dehydration}}$$

part of ricinoleic acid chain
(castor oil)

$$\text{CH}_2\text{CH}=\text{CHCH}=\text{CHCH}_2$$

conjugation produced by
dehydration of castor oil

Another drying oil of increasing importance is tall oil, so-called after the Swedish word for 'pine tree', from the wood of which it is extracted. After treating pine-wood pulp with a mixture of sodium hydroxide solution and sodium sulphide under pressure, the spent liquor is acidified. This yields crude tall oil which is a complex mixture of rosin and fatty acids, cyclic alcohols and other organic compounds. Volatile components are first removed ('stripped') by heating at about 250 °C under reduced pressure in a stripping tower. The residue is fractionally distilled, the various fractions being used for the production of detergents, alkyd resins, varnishes, paints and many other materials.

Three main types of surface coating films are prepared from drying oil bases—paints, varnishes and linoleum.

The use of drying oils as a liquid medium ('vehicle') for the manufacture of coloured paints is said to have originated in China. It was not used in Europe, however, until about the thirteenth century, although linseed oil was known in Roman times and was mentioned by Pliny. In classical Greece and Rome paint pigments were still mixed with materials such as gum arabic, egg white and gelatin. Varnishes were known to the Egyptians as early as 1000 BC, as revealed by their use in preserving mummy cases. In the absence of suitable solvents, resins were mixed in the molten state and applied hot with a palette knife. The use of drying oils for varnish making is first mentioned in a tract of the Greek scribe Aetius written in the sixth century, but the earliest recorded recipe for a heat blended linseed oil/resin varnish is to be found in an eighth-century manuscript possessed by Lucca Cathedral in Italy. No less than ten different resins mixed with about 14% of linseed oil are mentioned in the Lucca manuscript.

The manufacture of varnish is also described by the eleventh-century monk Theophilus in his *Diversarum Artium Schedula*. Linseed oil was heated in an iron pot and then stirred into a molten resin mixture. The whole was then heated, samples being withdrawn for testing at intervals, until the required degree of clarity was obtained. Soon after this in the reign of Edward I in 1283 the Paynter & Stayner's Guild was founded (*painters* worked on metal, timber, glass and plaster, *stainers* on fabrics such as canvas, linen and cloth). In 1290 a gallon (4·5 l) of linseed oil cost eight and a half new pence being imported from northern Europe via Hull. It is interesting to note that by 1316 Nicholas the Paynter was being employed in Exeter Cathedral at a wage of ten new pence per week—four times the wage of an agricultural labourer at that time.

The first modern varnishes containing driers and a solvent were not manufactured until the late eighteenth century—the first British factory being built in 1790. The addition of phenolic type synthetic resins to varnish in 1910 gave rise to the bakelite varnishes, followed by the oil-modified alkyd resin varnishes of the 1920's. Finally, just after the beginning of World War II, emulsion paints made their appearance, to be followed by the 'non-drip' thixotropic paints and polyurethane varnishes.

A paint has been defined as 'a liquid containing a suspended pigment to adhere to a solid surface and to protect and decorate it'. Paints ordinarily consist of powdered solid pigments dispersed in a liquid vehicle. The vehicle is usually a mixture of resins with sufficient added solvents to create the optimum consistency for brushing, spraying or dipping. Most paints, such as the oil paints, emulsion paints and distempers, are designed to dry at ordinary temperatures,

Type of paint	Million gallons (million litres in brackets)		Value (£ sterling)
Oil and/or synthetic-based (non-aqueous)	60	(270)	90 000
Oil and/or synthetic-based (aqueous)	1	(4·5)	1 600
Emulsion	16	(72)	22 000
Water paint	2	(9)	1 500
Cellulose-based	13	(58·5)	14 000
Varnishes, stains and lacquers	1·3	(5·85)	1 500
Marine	1	(4·5)	1 800
Bituminous	0·96	(4·32)	1 600

Production of paints and varnishes in the UK (1968)

but stoving paints or enamels have to be heated to obtain a hard surface film.

Oil paints contain both inorganic and organic pigments. The inorganic type are more commonly used because of their cheapness and fastness to light, and include materials such as Prussian Blue, red lead, zinc oxide, titanium dioxide, lead chromate, iron oxide and carbon black. Besides helping to obliterate the colour of the surface being painted, pigments strengthen the paint film. Cheap inert fillers

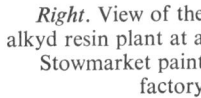

Above. A sample of paint being taken from one of a battery of fixed mixers

Right. View of the alkyd resin plant at a Stowmarket paint factory

such as whiting and calcium sulphate are often used to reduce the quantity of pigment required and for this reason are known as 'extenders'.

Solvents such as petroleum white spirit, solvent naphtha and turpentine are used as 'thinners' to reduce the viscosity of the paint during application and to clean paint-brushes after use. Turpentine is seldom used today because of its high cost. The 'vehicle' or liquid medium in which the pigment is suspended is usually a linseed oil mixture containing 'driers'.

Gloss paints which are required to have a very smooth shiny surface have an oil varnish or oil-modified alkyd resin varnish base. Priming paints contain suspensions of substances such as iron oxide, lead oxide, zinc chromate or aluminium dust in linseed oil, together with driers and thinners. Other special purpose paints include bituminous paints which are used to protect ironwork from corrosion, and anti-fouling compositions which are used to paint the hulls of ships. The latter contain salts of mercury or copper, which are slowly liberated in contact with sea water, and are toxic to aquatic organisms.

Emulsion paints contain a suspension of pigment in a stabilized aqueous emulsion of either a drying oil (oil-bound water paint) or a synthetic resin such as polyvinyl acetate with a plasticizer (plastic emulsion paint). Distempers consist of dispersions of whiting and coloured pigments with a glue or casein-based binding agent. Organic pigments such as coloured diazo compounds are often used in distempers. Stoving paints or enamels for metals are prepared by dispersing the coloured pigment in a partially cured synthetic resin. The most common types of resin used for this purpose are of the urea/formaldehyde, alkyd or epoxy type (see *High Polymers*, Book 1 in this series). After thinning, the enamel is sprayed on the metal surface to be painted which is then passed through a heated tunnel. This heat-cures the resin to give a hard glossy surface, which is widely used for such things as car bodies and domestic appliances.

Another type of resin-based paint requires a curing agent to be mixed with it just prior to application. The resin is cured by this means at room temperature, producing a tough, protective surface film. This is popularly called 'two-can' paint because the two interacting liquids have to be stored in separate containers, their 'pot life' after mixing being comparatively short, varying from a few minutes to a month or two, according to the type of resin system used. Acid-cured two-can paints based on U/F (urea/formaldehyde) resins have a long pot life of at least a month. They are mainly used for floor treatment and furniture finishing as the curing solution which contains sulphuric acid would affect metal surfaces. Mixtures of U/F resins and epoxy resins can be cured using primary and secondary amines such as diethylene triamine. They are suitable for metal surfaces but only have a pot life of a few hours. The pot life of epoxy-amine paints can be considerably extended, however, by using formulations which slowly release amine type curing agents on contact with atmospheric moisture. Other types of two-can system based upon polyurethanes and unsaturated polyesters have been replaced by more convenient one-can formulations which are gaining in popularity.

Non-convertible Systems

Although not oil-based, it is convenient to mention at this stage paints which do not 'dry' as the result of chemical action but by the evaporation of a solvent. These are known as non-convertible paints and are usually applied by spraying as they do not brush well, being very rapid drying. The most widely used non-convertible paints are the cellulose nitrate lacquers. Industrial cellulose nitrate (lacquer grade nitrocellulose) is explosively inflammable and is therefore supplied to the manufacturer 'damped' with an alcohol such as propanol or ethanol in which it is insoluble. Cellulose nitrate is only soluble in relatively expensive organic solvents such as ketones and esters and this inevitably raises the cost of the paint.

Although cellulose nitrate films produced by the evaporation of simple solutions are hard and quick drying, they are brittle and have poor adhesion and gloss. It is necessary therefore to add other ingredients to modify these characteristics. A plasticizer such as di-n-butyl phthalate is added to give the film pliability, and resins such as ester gum (rosin partially esterified with glycerol) or one of the alkyds are used to improve adhesion and brilliance.

A saving in solvent cost can be made by using highly viscous 'hot-spray' nitrocellulose lacquer designed for use at a temperature of about 70 °C. An additional advantage of these hot-spray lacquers is that the film produced is about twice as thick as that which can be obtained by using cold-spraying techniques.

Lacquers based on PVC (polyvinyl chloride) are attractive because of their high resistance to water and other chemical substances. Unfortunately the regular structure of the PVC molecule enables the polymer chains to lie close to each other and this permits considerable cross linking to occur between the polar chlorine atoms on one chain and the hydrogen atoms on another.

Formation of hydrogen bonds
between adjacent PVC chains

As a result of this ordered structure it is almost impossible to dissolve PVC at room temperature in any of the commercial solvents. This problem has been overcome by copolymerizing PVC and PVA (polyvinyl acetate). The PVA units along the polymer chain upset the regularity of the structure thus allowing penetration by the solvent molecules. The ratio of PVA units to PVC units in the polymer is usually adjusted to about 1:6. Copolymers of this type are soluble in ketones and esters, but retain most of the desirable characteristics of PVC such as resistance to chemical attack. The proportion of solvent used has to be carefully regulated to avoid the formation of irreversible gels on standing. Vinyl lacquers are used as chemically resistant paints for coating the insides of food tins, and for protecting metal surfaces from sea water corrosion. Films of lacquer are often applied by passing sheets of the material to be painted between rollers coated with the lacquer fluid (roller coating). Alternatively a dispersion of powdered PVC plasticizer and pigment is sprayed on the surface to be treated, which is then stoved at a temperature of about 170 °C (organosol coating).

Acrylic resin lacquers are used for motor car bodies, and possess a good durable gloss with excellent colour retention. Polymethyl-methacrylate is the hardest of the acrylic resins and is most commonly used.

Chlorinated rubber paint is ideal for producing a cheap chemically resistant surface at low cost. By treating natural rubber with chlorine gas a creamy coloured powder is produced, containing just over half its own weight of chlorine. This is dissolved in aromatic hydrocarbon mixtures, and after the addition of a filler and plasticizer can be applied by either spraying or brushing. It is specially useful for treating concrete surfaces, chemical plant and bathroom or kitchen ceilings which are exposed to excessive condensation.

Electropainting is now widely used industrially. Two alternative techniques are available.

Electrostatic spraying—the object to be sprayed is earthed and an electrostatic charge applied to the spray gun. The paint droplets leave the spray gun charged and are attracted to the earthed surface. This technique is especially effective when painting irregular surfaces with many crevices or involutions. 'Safe' voltages of up to 100 kV are used for hand spraying but automatic electrostatic spraying units have been constructed using voltages of up to 10 MV.

Electrocoating—with this technique the object to be coated is carried on a charged conveyor and dipped into a paint bath. Low voltages (50–150 V) are used at high amperage levels (1–5 kA) and the process resembles electroplating. The safety and efficiency of this method has led to its use in the motor car industry for painting vehicle bodies etc.

Varnishes

Varnishes are solutions of natural or synthetic resins blended with drying oils, with the exception of spirit varnishes (French polish) which are simple solutions of resin in methylated spirit. Some of the hard natural resins used for varnish making are obtained from the fossil remains of extinct varieties of trees, such as Congo copal, which is found buried in the tropical swamps of the Congo Basin. Other natural varnish resins include 'Dammar', 'East India' and rosin. These resins are insoluble in the drying oils used which include linseed, dehydrated castor oil and tung oil. To overcome this difficulty the resins are heated to about 320 °C and kept at this temperature for a short time ('gum-running'). This heat treatment partially decomposes the resins ('degradation') to give lower molecular weight compounds which are soluble in the drying oils. Synthetic resins such as phenoplasts, maleic resins, ester gums and polyurethane are commonly used in the preparation of varnishes.

The resin-oil mixtures are blended with a little driers and dissolved in a volatile solvent such as solvent naphtha (a petroleum distillate boiling between 120 ° and 200 °C) or turpentine. After application the solvent evaporates and the oils harden to give a hard, durable film. Reduction in the oil content ('short oil' varnish) gives a harder but rather brittle surface suitable for indoor use. 'Long oil' varnishes with a higher proportion of drying oil give a pliable film which is more resistant to outdoor weathering.

Alkyd resins can be modified by treatment with drying oils to give a product which is soluble in white spirit (a high boiling naphtha). Modified alkyd resin solutions of this kind only require the addition of driers to produce a serviceable varnish.

French polish is made by dissolving shellac (a resiny scale produced by the lac insect) or a phenolic resol in methylated spirit. Shellac varnish is air dried but the phenolic version requires stoving. Although once widely used for surface finishing furniture, French polishing has now been almost entirely replaced by the more durable and easily applied synthetic resin lacquers.

LINOLEUM AND OILCLOTH

Oxidized boiled linseed oil is the essential raw material used in the manufacture of linoleum. The original process involved the repeated dipping of cotton sheets ('scrims') into boiled linseed containing driers. Between dipping, the sheets were hung in sheds maintained at a temperature of between 35 ° and 45 °C until the oil had hardened. After several weeks a coating up to 25 mm thick could be produced.

In modern processes the oxidation of the boiled linseed oil is accelerated by allowing it to spray down through a warm air blast ('showering') and heating the thickened oil, while stirring vigorously and blowing air through the pasty mass. When the required stage of oxidation has been reached, resin is mixed with the oil paste to produce what is termed 'cement'. After maturing, the cement is kneaded with ground cork, pigments and other ingredients. The resulting dough is pressed on to a jute canvas backing using steam heated rollers (calenders) which produce a smooth surface. The sheets of linoleum are then matured in a hot room for some days before printing. Ordinary printing is unsatisfactory because the surface pattern soon wears off, therefore several methods of producing 'inlaid' linoleum have been used which give a 'right-through' pattern. Usually separate coloured pieces of linoleum are pressed together on the jute backing to give a mosaic; alternatively coloured cement is fed onto plain linoleum using special stencils and the colour-ed material pressed into it using heated rollers or hydraulic presses.

By brushing thickened linseed oil on a light cloth, thinner layers of oxidized oil can be produced. After printing, this is used as a flexible, waterproof covering for shelves and tables known as oilcloth. Lately, however, this type of covering has been superseded by pat-terned self-adhesive plastic sheeting.

Candles

Until the early part of the nineteenth century candles were made by covering wicks with mixtures of beeswax, tallow and spermaceti by pouring or dipping. After about 1830 a cheaper mixture of palmitic and stearic acids ('stearine') was used. This was obtained by hydro-lysing selected animal fats ('fat-splitting'). These stearine candles were rather brittle, however, and later paraffin wax was used to make them more pliable. The paraffin wax content has gradually been increased, until at the present time it has largely replaced the use of solid fatty acids. The latter are still produced by splitting animal fats such as tallow and then crystallizing out the solid fraction. After pressing to remove the liquid fraction (oleine) the purified solid is sold com-mercially as stearic acid. In the Emersol process the fatty acid mixture is dissolved in a 1:9 water/methanol mixture before chilling and filtering off the solid stearine.

Removing a clamp of finished candles from a machine at Price's Belmont works, Battersea, London

Other Uses of Oleochemicals

In addition to the use of fats and fatty acid esters in cosmetics and for carrying the active ingredients in pharmaceutical preparations such as ointments and liniments, certain fats and their derivatives have direct medicinal applications. Croton oil and castor oil are effective purgatives, and chaulmoogra oil and hydnocarpic oil and their methyl esters are used to treat leprosy. Fish liver oils can act as food additives on account of their high vitamin A and D content.

After the tanning of leather, the hides are impregnated with fats or fatty emulsions to give an oil content of from 5 to 20%, and thus improve their pliability. Chamois leather is tanned using certain unsaturated fish oils with which the skins are impregnated before being hung up in warm rooms to allow the tanning process to take place.

Palm oil and tallow are applied in the tin-plating industry to cover the exposed surface of the tin to prevent oxidation and to lubricate the flow of tin over the metal sheet during the coating process.

Apart from the soluble soaps described in detail in the following section, a number of insoluble metal stearates are used as lubricants:

calcium and zinc stearates as mould lubricants in the plastics industry; magnesium stearate in talcum and face powders; aluminium stearate as a lubricating oil additive to increase viscosity, and also for the waterproofing of leather and fabrics.

An ester of cetyl alcohol (cetyl caprylate—Estol 22) is extensively applied in the weaving industry as a yarn lubricant. Although originally intended for the processing of linen it has also been used successfully for cotton, rayon, nylon, tri-acetate, polyester and acrylic fibres. Similar compounds have been used for lubricating loom control cords and shuttles.

In 1955 the Commonwealth Scientific and Industrial Research Association in Melbourne carried out a series of experiments to conserve natural water reserves in Australia using monomolecular films of long chain alcohols. Assistance in this project was sought from the English oleo-chemical firm of Price's (Bromborough). As a result small beads of cetyl alcohol (hexadecanol) were produced by subjecting a liquid jet of the molten material to sonic vibration. The beads were submerged below the water surface in wire cages and produced water layers ('rafts') of about two nanometres (2×10^{-9} metres) thickness. This allowed satisfactory penetration by oxygen and rain droplets but inhibited normal evaporation. Further tests carried out in Loch Laggan (Scotland) in 1960 showed that floating dispensers containing a solution of cetyl and stearyl alcohols in kerosine could also be used for water conservation.

Chapter 3

Detergents and Polishes

Detergents and polishes provide interesting examples of the commercial application of fats and waxes. Both are universally used for the treatment of surfaces which require to be cleaned or protected. Both are of special interest to the housewife who for generations past has made use of these materials to keep her home and its contents bright and clean. In the last half century the traditional pre-eminence of soap as a cleansing agent has been challenged by the arrival of many new soapless detergents based upon petroleum chemicals. The use of additives such as 'fluorescers', enzymes, bleaches and soil dispersing agents has revolutionized laundering. Manufacture of detergents in new forms and by continuous processing instead of batch processing is another feature of this giant modern industry.

The rise of the petrochemical industry has also made available new solvents and paraffin wax, which are important constituents of many of the multitude of polishes produced today for the treatment of a variety of surfaces, from furniture and floors to metals and motor cars. Here again the addition of new materials such as silicones has enabled big advances to be made in the formulation of polishes. The increasing use of emulsions and soft pastes, and dispensers such as the aerosol can have also helped to reduce the work involved in polishing while producing a high protective gloss.

DETERGENTS

Soap is the oldest and still the most widely used of all detergents (L. *tergere*—to wipe clean). Its origin is obscure but it is certain that soap-like materials were being used in ancient Mesopotamia as early as 2500 BC. In the Old Testament (Jeremiah 2: 22) there appears the phrase: 'For though thou wash thee with nitre, and take thee much soap, yet thine iniquity is marked before me, saith the Lord God.'

It is believed the 'soap' referred to was probably a lathering paste of Fuller's earth and impure soda mixed with urine. Pastes of this kind were in common use by the Romans for cleaning clothes and pots of the material were found in the ruins of Pompeii. The Roman chronicler Pliny the Elder, who perished in the blazing inferno of Pompeii in AD 79 mentions the use of 'sapo' prepared by boiling tallow from goats with beech-wood ashes. It is clear from what he says, however, that this was not used for washing but as a kind of primitive hair cream. Almost a hundred years elapsed before Galen the Greek physician mentioned a soap for washing the body.

The origin of the word 'sapo' is Sapo Hill on the outskirts of Rome, the site of many Roman temples to heathen gods. Washerwomen used the clayey earth on the banks of the River Tiber immediately below this hill because of its remarkable cleansing properties. It has been suggested that the animal fat from burnt sacrifices was saponified by reaction with the hot wood ashes of the sacrificial fires to form a crude soap. This was thrown out on the hillside when the temples were cleaned and washed down to the river bank by rain water, to be absorbed by the surface clay.

References to a material for washing cloth appear among the writings of the ancient Egyptians who made use of natural deposits of crude soda (nitre) in the Nile Valley, and castor oil. It is of interest that certain primitive peoples still use mixtures of wood ash and natural oils (cold soap) for washing, as well as lathering pastes and powders made from roots, nuts and berries containing the material saponin.

Very gradually the knowledge of soap making spread from the Mediterranean to other countries, especially France, Italy and Spain. These had supplies of alkaline barilla (ash obtained by burning sea plants), and olive oil which was used in place of animal fat. The French word for soap (*savon*) is named after Savana, an important soap-making city of the twelfth century. In the same way, the term Castile soap is a reminder of the fine white luxury soap of that time which was produced in Venice and Castile.

The manufacture of soap in Britain appears to have begun shortly before the Norman invasion in the eleventh century, although some reference to soap making is made in the Saxon *Leechdoms* or medical books almost two hundred years before this. By 1180 Bristol

5

had become an important soap-manufacturing centre and 12 years later, Richard of Devizes—a travelling monk from Winchester—wrote scathingly of the number of soap makers in the city and the foul stenches they produced. The name 'Bristol' became associated with a coarse grey soap which was still used under that name in Holland until quite recently.

By the time of Queen Elizabeth I soap was being made in most of the large English cities such as Coventry, Hull and York, and there is mention of 'sopehouses' being set up in London at Blackfriars, Cheapside, Grasse Street (Gracechurch St), Bermondsey and Southwark. Three main types of soap seem to have been available at this time. A coarse scrubbing soap made from whale oil, the luxury toilet soap from Castile, and the grey Bristol soap made from tallow and wood ash. Even the cheaper soap cost about 1d a pound however ($\frac{1}{2}$ p/450 g—about 12$\frac{1}{2}$ p today) and washing oneself was regarded as rather a luxury. A friend of Queen Elizabeth boasted that she 'hath a bath every three months whether she needeth it or no'. Even the great Dr Johnson thought that total immersion of the body for washing purposes was an unnecessary risk, and counselled one of his friends who thought of trying the idea to 'let well alone and be content'. It is also interesting to find Samuel Pepys in 1659 choosing 'a barrel of soap to be carried to Mistress Ann', an expensive gift for one of his lady friends, the Lady Ann Montagu.

Both in England and in France the high cost of soap was partly due to the many petty restrictions and taxes which its manufacture seemed to attract from the legislators. In 1634 Charles I issued a proclamation which gave the London soap makers an almost complete monopoly of production, killing the Bristol trade, and in 1649 a tax of 4s (20p) a barrel was imposed doubling the retail cost. To prevent illicit production during darkness, the soap boiling pans were fitted with wooden lids which were locked at night by excise officers.

A great technical advance at this time was the discovery of the 'salting out' technique which facilitated the separation of soap from the spent lye. An account of this is given in a recipe book compiled by the French physician Turquet de Mayerne. Describing the processing of a soap mixture after boiling he says: 'Then strew in salt, and if the lye part from the tallow, dip in a skimmer and hold the one

An eighteenth-century soap pan room

side downwards—you shall perceive the soap run down the skimmer like flakes of snow.'

Adoption of the new technique was slow, however, partly because of the crippling tax on salt and also because of the great secrecy that surrounded soap-making practice. Little scientific method was employed and many of the processes were very crude. For instance, it was recommended that the strength of the alkali liquor (lye), be determined by observing the behaviour of a new laid egg floated upon its surface!

Following Scheele's discovery of glycerol, however, in 1779, there were two further advances which placed soap making on a firmer scientific footing. In 1789 the young Frenchman Nicolas Leblanc, family doctor to the Duke of Orleans, devised a process for the cheap production of soda from brine. This replaced barilla, and 'Dantzig' ashes, previously imported into Britain from the Baltic as sources of alkaline lye. Strangely enough the new 'soda ash' of Leblanc was at first rejected by the soap makers and James Muspratt, who produced it in England, had to give away hundreds of tonnes free of charge to popularize it. The chemical processes involved in soap making were clarified as a result of the brilliant work of Michel Chevreul, who in 1816 made a systematic study of the properties and structure of fats. The major British soap companies were being founded at about this time, and the Great Exhibition of 1851 included a vast display of soaps in a variety of forms and colours, which reflected the thriving state of the industry. W. H. Lever was the first to sell wrapped and branded soap in Bolton during the year 1884, followed by the introduction of soap 'flakes' in 1899. Improvements had also been made in the design of soap-making plant, the first patent for a soap kettle with a capacity of 60 barrels having been taken out by David Ramsey in 1636, followed in 1721 by the patenting of the first steam heated soap kettle.

By the latter half of the nineteenth century special machinery for the making and blending of toilet soap in moulded pieces was already being used. Meanwhile soap kettles grew in size and improved in design, the modern variety holding up to 150 tonnes of ingredients at a time. Although soap is still mainly manufactured in kettles, a recent innovation which is gradually replacing this older batch method is the continuous soap-making process, introduced into Britain in 1940 by Thomas Hedley and Co.

During the eighteenth century, a few years after Chevreul's classic investigations into the chemistry of fats, another discovery had been made in France which was destined to revolutionize the detergent industry. In 1831 Edmond Frémy noticed that on treating olive oil with concentrated sulphuric acid and neutralizing the resulting product he obtained a liquid with a soapy feel which lathered in aqueous solution. The potentialities of these new soap-like materials as dyeing assistants were noted by the German dyemaster Runge in 1834. As a result a number of other plant oils and animal fats were sulphated but although a few, such as sulphated castor oil ('Turkey Red Oil') produced by Crum in Glasgow in 1875, found use as wetting agents in the dyeing industry, they were disappointing as laundering and washing agents. Their one important advantage, however, was the absence of scum production in hard water, which had always been one of the drawbacks of soap.

Two factors stimulated further research into the new type of detergents. One was the shortage of fats and oils occasioned by the increased demand and the rise of the margarine industry; the other was the ever expanding field of petroleum chemicals which was opening up exciting new possibilities in the way of raw materials. Just before World War I the Belgian chemist Reychler produced some interesting liquids which had the surface active properties of soaps ('surfactants'), by sulphating some of the higher alcohols such as cetyl alcohol. Three years later in 1916 the German chemical firm of Badische Anilin und Soda-Fabrik A.G. produced a sulphonated substance derived from naphthalene and iso-propyl alcohol. This was marketed as 'Nekel A' and as it possessed both aromatic and aliphatic components it was termed an alkyl aryl sulphonate. Meanwhile research along similar lines was being carried out in Britain and the USA and by the 1920's it was obvious that the number of synthetic surfactants which could be synthesized was enormous. The production of soapless detergents on a commercial scale rapidly gained momentum and soon after the end of World War II, spray dried detergent powders appeared in Britain which were based upon sulphated alcohols and alkyl aryl sulphonates.

A few years later liquid washing-up detergents appeared on the market and soap was soon being replaced for all purposes except toilet use and commercially in launderies using softened water. At the

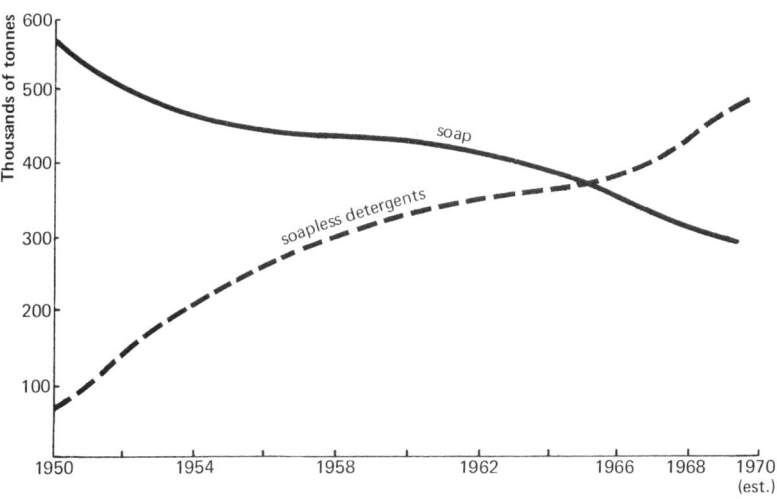

Fig. 3.1 Production of detergents in the UK

present time the consumption of synthetic detergents ('syndets') in the USA and several European countries exceeds that of soap.

In the USA synthetic detergents account for over 80% of all detergents used and are even incorporated in toilet soaps. West Germany and France also use a high percentage of syndets but in the UK there is still a fair market for soap-based washing powders. In the developing countries of the African, Asian and South American continents

1967	Total usage of detergents (kg/head (pounds in brackets))	Synthetic detergent %
West Germany	13·10 (29·1)	79·8
France	12·00 (26·8)	75·0
Italy	9·75 (21·6)	64·4
Portugal	9·25 (20·5)	16·6
Belgium	13·60 (30·3)	69·0
Netherlands	13·60 (30·3)	67·5
United Kingdom	12·10 (27·0)	51·2
USA	16·20 (36·0)	81·7

the consumption of soapless detergents is small but likely to develop rapidly within the next few years.

The rapid growth of the soapless detergents industry resulted in a fall in the production of glycerol, traditionally regarded as a by-product of soap making. This deficit has now been more than made up by the production of glycerol from petroleum chemicals (see *Fuels, Explosives and Dyestuffs*, Book 3 in this series).

An interesting development in the design of new detergent mixtures has been the use of auxiliary materials to enhance the action of the detergent base ('builders'). The addition of builders to soap was suggested by Lemery in 1727 and by the end of the eighteenth century materials such as rosin, talc, starch and borax were commonly used as fillers. By the beginning of the present century this range had been extensively added to in the preparation of detergents for laundering purposes. Two builders of special interest which have been used in post-war washing powders are sodium carboxymethylcellulose, which effectively prevents the redeposition of dirt on laundered materials, and 'optical bleaches' or 'fluorescers' which brighten white fabrics by causing them to fluoresce in daylight.

Other builders commonly used in modern detergent powders are water softeners such as phosphates, sodium perborate which liberates

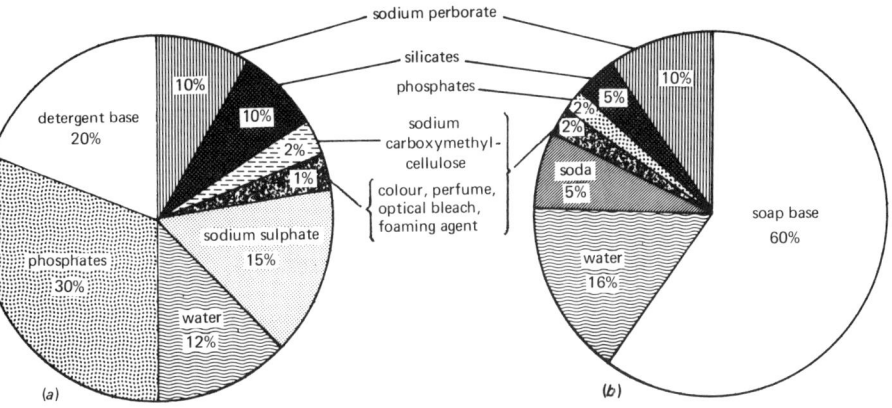

Fig. 3.2   Typical analysis of (a) *a soapless detergent powder and* (b) *a soap powder*

oxygen in hot water to assist bleaching, and bulking powders such as sodium sulphate which also improve the physical characteristics of the powder. Silicates are common additives which assist detergent action, keep the powder crisp, and prevent staining of aluminium articles. Foam stabilizers are often used to promote good lathering characteristics.

More recently enzymes have been included in so-called 'biological' detergents to digest protein-based stains such as blood, sweat and egg, which are not satisfactorily removed by traditional detergents. These proteases are derived from various strains of *Bacillus subtilis* and are stable enough to withstand the manufacturing process without being inactivated. About 1–1·5% of the active enzyme is included in the detergent formulation and overnight soaking is recommended before laundering. This is necessary because the enzyme is destroyed on heating. A good deal of research has been carried out to investigate the effect of detergent ingredients such as phosphates on protease activity.

Properties of Detergents

The advances made in the detergents field during this century have been the result of intensive research into the structure of detergent molecules and their mode of action. By the turn of the century, work by Krafft and others had established that soap ionized in solution and also contained colloidal particles of some kind. The term 'colloidal electrolyte' was used by Duclaux in 1907 to describe this type of compound. It had been noted by McBain that fatty acid salts with more than seven carbon atoms in the hydrocarbon chain showed a tendency to clump together in solution to form particles of colloidal dimensions. In 1913 he suggested that the properties of solutions of soaps and other surface active materials could be explained by the formation of minute wafer-like molecular rafts. These were envisaged as being tightly packed layers one molecule thick, having

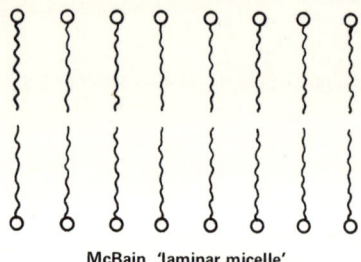

McBain 'laminar micelle'

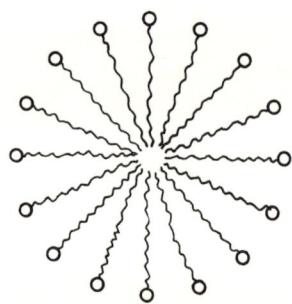

Hartley 'spherical micelle'

O 'head' unit (substituted carboxyl group)

~~~~~ 'tail' unit (hydrocarbon chain)

Fig. 3.3

hydrocarbon chains on one side and substituted carboxyl units on the other. McBain further considered that these layers clumped together in pairs to form a molecular sandwich with the hydrocarbon tails as the 'meat' in the centre. These orientated molecular sandwiches were called 'laminar micelles'. Later Hartley suggested that 50 or more surfactant molecules could arrange themselves into a sphere with all the hydrocarbon tails pointing inwards toward the centre of the sphere, producing a surface layer of the ionizing substituted carboxyl group. The formation of 'spherical micelles' seems to be more likely at lower concentrations, but with highly concentrated solutions there is fairly strong evidence that laminar micelles exist.

It is interesting to note that micelle formation only occurs when the concentration reaches a certain value known as the critical micelle concentration (c.m.c.) and then takes place suddenly and rapidly. This value decreases with hydrocarbon chain length, and suggests that the non-formation of micelles in the case of fatty acid salts with very short hydrocarbon chains is due to a very high c.m.c. which is virtually unattainable. On the other hand, it must be remembered that the solubility of a surfactant decreases as its hydrocarbon tail increases in length. These two factors have to be balanced when designing a commercial surfactant material.

Although most of the early work on the theory of surface activity was carried out using solutions of soaps, it soon became apparent that all surfactant molecules had a similar structural pattern which was responsible for their special properties. In the simpler cases a water-loving (hydrophilic) 'head' unit is attached to a hydrocarbon 'tail' which is water-hating (hydrophobic). Great diversity is possible in the number and nature of the heads and tails, giving an infinite variety of surfactant materials.

Molecules which possess both water-loving and water-hating components in this way are termed amphipathic. The 'wetting' and detergent properties of the surfactants are both due to their amphipathic nature. Because of their hydrophobic behaviour there is a tendency to expel the hydrocarbon chains from solution to prevent contact with water as far as possible. At low concentrations this can only be achieved by pushing the surfactant molecules to the surface where the hydrocarbon tails are forced out of the water approxi-

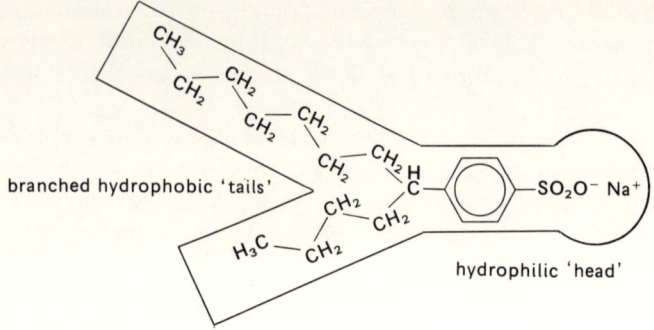

hydrophobic (water-hating)   hydrophilic (water-loving)
'tail' unit   'head' unit

(*a*) Diagrammatic representation of a soap molecule (sodium stearate)

(*b*) Diagrammatic representation of a synthetic detergent molecule
(alkyl aryl sulphonate)

mately parallel to each other, leaving the hydrophilic heads still dissolved. This accumulation of surfactant molecules at the water surface lowers the surface tension, which in turn allows water droplets to flatten out and 'wet' surfaces with which they are in contact. Substances such as sugar and salt which are not surface active tend to increase surface tension.

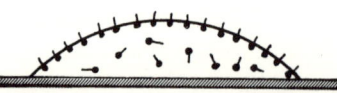

Spherical drop of water unable to wet the surface with which it is in contact

Surface tension of water drop reduced by the addition of a surfactant — this has allowed the drop to spread and wet the surface

Fig. 3.4

If oil is present, the surfactant molecules concentrate along the oil/water boundary in a similar fashion. The lowering of tension between the two liquids is very marked in this case because the hydrocarbon tails are able to dissolve in the oil. This partly explains the detergent action of surfactants and their use as emulsifying agents.

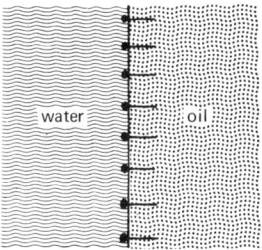

Fig. 3.5   Arrangement of surfactant molecules at an oil/water interface

At concentrations above the c.m.c. a further mechanism becomes available to reduce the contact between the water and the hydrophobic tails as micelles are formed, thus allowing the hydrocarbon chains to be tucked away from the water. Greasy soil and dirt are able to enter these micelles and become 'solubilized', another important factor in detergent action. The capacity of the micelles to envelop dirt in this way is quite remarkable and can double their volume.

Although most detergent solutions will produce a foam, many good foaming agents such as the naturally occurring saponins are useless as detergents. There appears to be no direct correlation between foaming ability and detergent power, although the housewife is usually still impressed by a copious lather and manufacturers often incorporate a foaming agent into washing powders as a result. It is interesting to note that this is often an embarrassment in washing machines and has led to the production of low level lathering products (e.g. 'Pat', 'Cheer').

Although large numbers of surfactants have been prepared, and the possible variations on the basic molecular pattern described above seem endless, only a few have become commercially important and these are classified according to their behaviour on solution in water. The most important class of surfactants ionize in water to produce two or more ions. If the ionized tail group carries a positive electrical charge the surfactant is said to be cationic. By far the largest group, however, are the anionics which ionize on solution to

Fats, Oils and Waxes

produce a negatively charged tail. The soaps and 95% of all soapless detergents come into this category.

Next in order of importance to the anionic surfactants are the non-ionics, which do not ionize in solution. These are usually prepared by reacting long chain fatty alcohols or fatty acids with ethylene oxide. Non-ionics have good detergent and emulsifying properties but are poor foamers.

The cationics are quaternary ammonium compounds which can be thought of as substituted ammonium chloride derivatives, the hydrogen atoms being replaced by organic groups, one of which acts as the hydrocarbon tail.

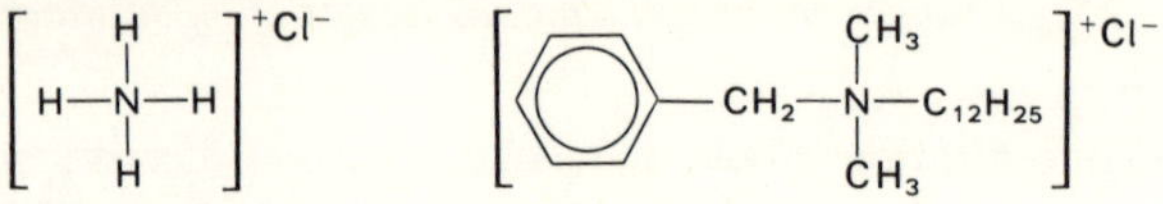

(*a*) ammonium chloride

(*b*) lauryl dimethylbenzyl ammonium chloride (a cationic surfactant)

Although poor detergents, the cationics are useful because of their powerful germicidal properties.

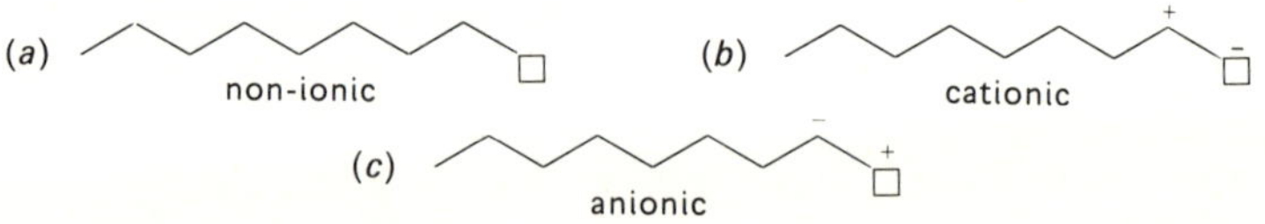

(*a*) non-ionic

(*b*) cationic

(*c*) anionic

A small group of surfactants which have only found limited commercial application to date possess both anionic and cationic properties. Molecules of this type are known as amphoteric and exhibit anionic characteristics when the solution is alkaline and cationic characteristics under acidic conditions. The pH at which both cationic and anionic properties are exactly balanced is known as the iso-electric point. All long chain amino acids behave as amphoteric surfactants.

Detergent Action

Having considered the structure and properties of surface active materials in general, we are now in a position to consider in more detail the action of soaps and soapless detergents. The function of these cleansing agents is to remove particulate and greasy dirt from surfaces such as skin, fabrics, flooring and crockery. To be of use a detergent must be effective in removing a wide range of soils such as house dust, soot, perspiration, food stains, blood, cigarette ash, cosmetics, cooking fat, fruit juice and ink. For this to be possible three fundamental requirements must be satisfied. The soil must be wetted, dislodged, and finally kept suspended until it can be rinsed away.

Wetting and penetration of the soiled material is promoted by reduction of surface tension due to the migration of the detergent molecules into the water surface, as described above. The detergent molecules are also attracted to the particles of dirt present on the soiled surface. Greasy particles attract the hydrocarbon 'tails' of the detergent which dissolve, leaving the water-loving heads on the outside. Ultimately the greasy soil is completely coated with a layer of detergent molecules which allows it to be lifted away from the surface being cleaned, helped by agitation of the liquid either by mechanical means or by the boiling action of the water. Non-greasy dirt is covered by an envelope of detergent molecules in the same way but in this case it is the 'heads' which are attracted to the dirt even more strongly than to the surrounding water. This in effect produces a greasy film on the surface of the dirt due to the layer of hydrocarbon tails present, and this is covered by a second layer of detergent molecules enabling the particles to be dislodged as before. The engulfing of soil particles in detergent micelles in this way prevents coalescence, thus helping to keep them in suspension.

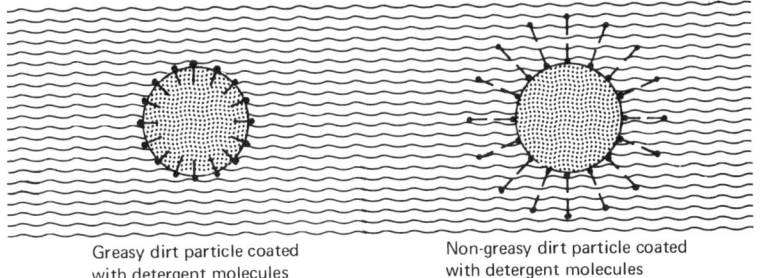

Greasy dirt particle coated
with detergent molecules

Non-greasy dirt particle coated
with detergent molecules

Fig. 3.6

$$2\ C_{17}H_{35}COONa + H_2O \rightleftharpoons$$

sodium stearate

$$
\begin{array}{c}
\quad\quad\quad \overset{|}{\underset{|}{C_{17}H_{35}C}}\!\!\diagdown^{\text{OH}} \\
\quad\quad\quad O \quad\quad Na^+ \\
\quad\quad\quad + \\
\quad\quad\quad O \quad\quad OH^- \\
C_{17}H_{35}C\diagdown_{OH} \\
\end{array}
$$

part of soap complex

In addition soaps hydrolyse to a slight extent in water to give minute amounts of free sodium or potassium hydroxide. This is able to saponify greasy soil, and is also partially responsible for the slippery feel of soap and for the stinging sensation when soapy water enters the eyes. Under acid conditions soap is rapidly hydrolysed to produce free fatty acids. This creates difficulties when soap baths are used during certain dyeing processes.

One of the drawbacks to the use of soap as a detergent is that hardness in water caused by the presence of metal salts in solution produces an insoluble sticky grey scum of metal soaps. This not only hinders the detergent action of the soap, but discolours and 'hardens' fabrics being washed. This difficulty can be partly overcome by the addition of builders such as phosphates, but it puts the soaps at a disadvantage when compared to the soapless detergents which are

$$C_{17}H_{35}COONa \xrightarrow{\text{aqueous solution}} C_{17}H_{35}COO^- + Na^+ \xrightarrow[\text{(hard water)}]{Ca(HCO_3)_2}$$

sodium stearate   ionized soap

$$
\begin{array}{c}
C_{17}H_{35}COO\diagdown \\
\quad\quad\quad\quad\quad Ca + NaHCO_3 \\
C_{17}H_{35}COO\diagup \\
\end{array}
$$

soap 'scum'

relatively unaffected by the presence of metal ions in solution. This partly explains the decline in the use of soap powders for home laundering since World War II.

Soap Manufacture

The most important soaps are the sodium and potassium salts of fatty acids containing from 12 to 18 carbon atoms. The harder sodium soaps are commonly used, the potassium soaps which form soft gels being preferred for special purposes where rapid solution is required. The physical properties of soaps depend largely on the nature of the fats used in their preparation. Mixtures of fats and oils are therefore blended to produce a soap having properties best suited for a particular use. Generally speaking, highly saturated fats of high molecular weight such as tallow give hard soaps which dissolve slowly but give a stable lather. Low molecular weight fats such as coconut oil, or those which are highly unsaturated such as groundnut oil, give soft soaps which dissolve quickly. Soaps made from low molecular weight fats also yield more readily dispersible scums in hard water.

The pH of a soap solution also depends to some extent upon the fats used in the preparation of the soap, the higher molecular weight fats such as tallow producing soaps with a solution pH as high as 10·8. Strangely, tallow soaps are milder to the skin than those produced from fats of lower molecular weight which give solutions with a pH of 9 to 10.

Soaps are most commonly produced by the saponification of fat blends with an alkaline solution. Increasing amounts, however, are made by the direct neutralization of fatty acids obtained from fat-splitting. During the traditional saponification process the fats are hydrolysed to fatty acids and glycerol, the acids reacting with the alkali present to form soap.

$$
\begin{array}{c}
\text{CH}_2\text{OCO} \diagdown\!\diagup\!\diagdown\!\diagup \\
\text{CHOCO} \diagdown\!\diagup\!\diagdown\!\diagup \\
\text{CH}_2\text{OCO} \diagdown\!\diagup\!\diagdown\!\diagup
\end{array}
\xrightarrow[\text{100-120 °C}]{\text{NaOH}}
\begin{array}{c}
\text{CH}_2\text{OH} \\
\text{CHOH} \\
\text{CH}_2\text{OH}
\end{array}
+ 3 \diagup\!\diagdown\!\diagup\!\diagdown \text{COONa}
$$

$$\text{fat (triglyceride)} \qquad\qquad \text{glycerol} \qquad\qquad\qquad \underset{\text{soap}}{}$$

$$(\diagup\!\diagdown\!\diagup\!\diagdown \text{ represents a hydrocarbon chain of } C_{12} \longrightarrow C_{18})$$

In the older 'boiling' process saponification is carried out in large cylindrical or rectangular steel vessels known as soap pans or kettles, which are usually open at the top and have a capacity varying from a few tonnes to as much as 150 tonnes. The lower part of the pan is funnel-shaped to allow easy drainage and contains a system of steam heating coils which can be either open or closed. Molten fat and a weak solution of sodium hydroxide ('lye') are simultaneously pumped into the pan and steam is admitted to boil the mixture. As saponification proceeds heat is generated, the reaction being exothermic, and stronger lye is added. The concentration of alkali must be sufficiently high to prevent the soap 'lumping', but not high enough to throw it out of solution. Boiling is continued until the greasy nature of the mix has disappeared and saponification is complete.

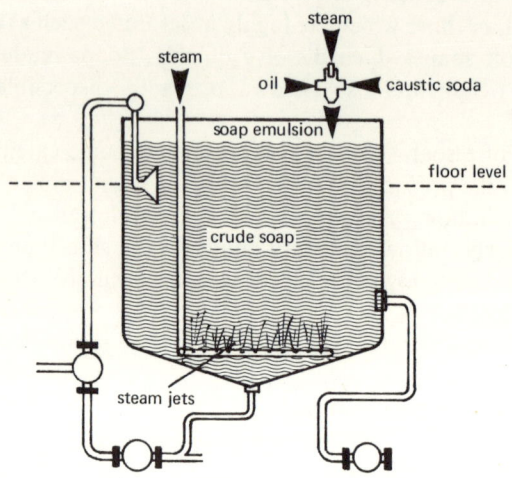

Fig. 3.7   Soap pan showing initial saponification

The second stage of soap making involves the separation of the soap and glycerol, a process known as 'graining' or 'salting out'. Use is made of the fact that soap is insoluble in concentrated salt solution, while glycerol is readily soluble. Solid salt or brine is added to the mix which is then boiled and allowed to settle. The soap is thrown out of solution as 'curd', and being of lower density than the glycerol/brine mixture, rises to the surface. The spent lye, which also contains glycerol, salt and dirt is then drawn off from the bottom of the pan and pumped to the glycerol recovery plant.

A number of washing operations are carried out to reduce the glycerol content of the soap and to remove impurities. It is successively dissolved ('closed') in water and after boiling for a short time is salted out, the lye being removed after settling. After each of these brine washings the lye is collected and evaporated to recover

A pan of soap being 'fitted'

the salt and glycerol present. The soap is then 'strengthened' by boiling with a strong alkaline solution to complete saponification, and after settling out as before, is 'fitted' or 'pitched'. The lye from the strengthening process is still strongly alkaline and after running off is added to the caustic soda used for the initial saponification process.

Fitting the soap involves adding water, while boiling, until the soap loses its greasy appearance and becomes smooth. The mix is then left for two to ten days, depending upon the size of the pan and type of soap being made, and slowly separates out into three layers. The upper layer is a high quality soap known as 'neat soap' which contains about 30% water, and is covered by a thin frothy curd called 'fob'. While it is still liquid the warm neat soap is pumped away using a skimmer pipe, which can be raised or lowered inside the pan on a swivel joint. Below the neat soap is a thinner dark liquid containing about 30–40% soap, termed nigre (L. *niger*—black), which is worked up into a lower grade soap. At the bottom of the pan is a layer known as 'pitch liquor' or 'nigre lye', which contains salt, alkali liquor and traces of glycerol.

In large soap plants with a number of pans it is economical to use a countercurrent process. To achieve this, a battery of pans is used

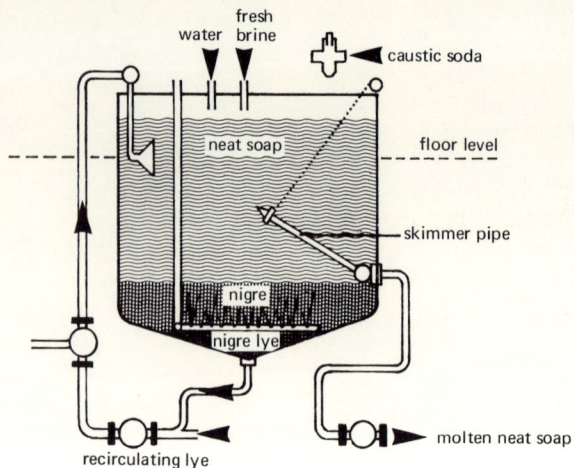

Fig. 3.8   Soap pan showing contents after fitting

containing soap in successive stages of manufacture, the washings from each pan being pumped into a following batch. In this way the lye moves progressively from the last wash to the first, running countercurrent to the curd which is progressively purified. This technique is cheaper and more efficient, enabling a more complete recovery of salt and glycerol and less wastage of alkali.

The manufacture of soft coconut oil or potassium soaps, and special superfatted or transparent soaps cannot be carried out by the boiling process. This is because of their greater solubility in water which prevents them being salted out. In this case a 'cold' process is used and the glycerol is not recovered but remains in the soap. Saponification is carried out by heating the stirred oil and alkali blend in a steam heated vessel called a 'crutcher'. This odd name is derived from the early stirrers used which were shaped like a wooden crutch. The hot mixture is then run into frames where saponification is completed.

Continuous saponification—*above:* a batch of centrifugal extractors for the removal of glycerol

right: a fitting column for the removal of free fatty acids

The manufacture of soap in batches is wasteful of time, equipment and space, and a great deal of research has gone into the development of continuous soap-making processes. One method involves the continuous catalytic splitting of fats with water in an autoclave at high temperature (230 °C) and pressure 4·13 MN/m^2 (600 lb/in^2). The fatty acids produced float on the water surface and are drawn off continuously, glycerol being extracted from the bottom of the autoclave in the same way. The fatty acids are then purified by distillation and neutralized directly with alkali to form soap.

In the modern continuous centrifuge process saponification can be carried out in about 15 minutes, compared to the hours required for the open pan method. This is achieved by using a closed vessel and reacting the fat/alkali mixture at elevated temperature and pressure. This not only has the advantage of speed but is economical of space, heat and manpower. It also permits automatic monitoring of the product, viscosity being controlled continuously by adjustment of the alkali feed. After cooling to prevent the pressurized product from boiling, the soap is washed and salted out, separation of the soap and lye layers being carried out by centrifugation. After filtering and re-centrifuging the soap is ready for shaping.

Another continuous process involves the emulsification of a blend of hot fat and alkali which is then pumped at high pressure through a heat exchanger maintained at 280 °C. After initial saponification has taken place the hot liquor is sprayed into a vacuum chamber where the glycerol and water evaporate ('flash off'). The remaining soap is processed by closing, salting out and fitting in the usual way.

The finished soap has additives such as rosin, colour, builders and germicides stirred into it during the crutching process. Traditionally the crutched soap is then poured into open-topped iron moulds each holding about 750 kg (15 cwt). After allowing about a week to elapse the cool matured soap is cut into small bars using steel wire cutters.

Here again continuous finishing processes are gradually replacing the older methods. In the 'Mazzoni' process molten soap is partially dried and cooled by spraying from the top of a vacuum chamber. The soap is scraped from the walls of the spray drier and passed under vacuum through two refiners or plodders which knead the soap and incorporate perfume. After extrusion and cutting into tablets the

Toilet soap being extruded during the refining process

Soap flakes being removed from the chilling roll

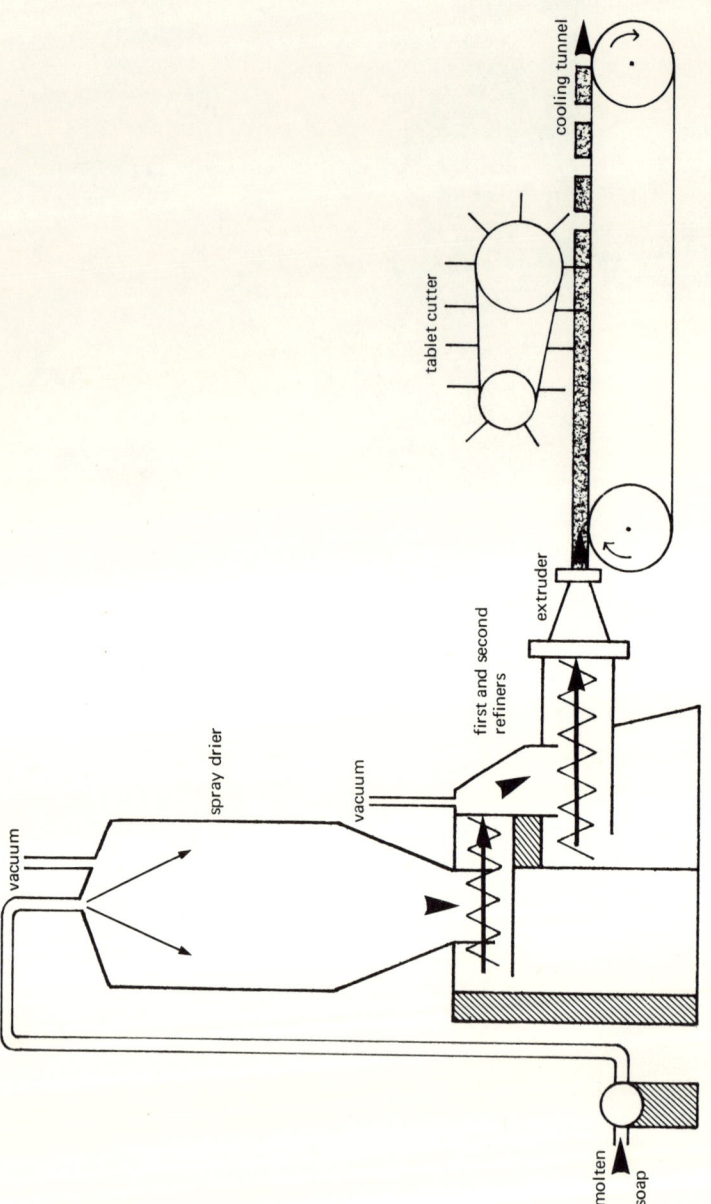

Fig. 3.9 Manufacture of bar soap

soap is conditioned by passing it through a cold air tunnel before stamping.

Toilet soap has a lower moisture content than household soap. This used to be achieved by spraying the hot soap on water cooled cylinders from which it was scraped in the form of thin shreds. The shredded soap was then dried by passing through a hot air tunnel—a process first used by Frères and Co. in 1891—before being rolled and finally extruded as a continuous bar. Modern processes use vacuum spray drying techniques as above, the soap being passed through two refiners, from the second of which it is extruded in the form of short cylindrical 'noodles'. After passing through a mixer which adds colour and perfume, the soap passes through a second pair of vacuum refiners to reduce the water content still further. The toilet soap is then extruded as a continuous solid bar which is chopped into tablets.

Toilet soaps containing synthetic detergents have been marketed since the end of World War II in the USA although combination soap/detergent bars did not become widely used until the late 1950's. The object of these soaps was to prevent the formation of scum and bath tub 'tide marks' produced by lime soap curds. Most formulations contain about 15% of a synthetic detergent which is commonly a long chain alkyl or aryl sulphate. One of the problems of manu-

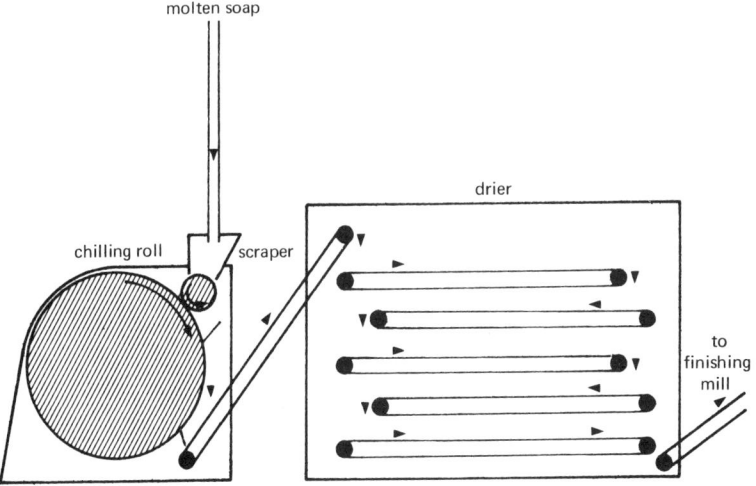

Fig. 3.10   Manufacture of soap flakes

facturing combination soaps is their tendency to 'slime' over and smear during use. Although almost unknown in this country it seems likely that this type of soap will be introduced shortly. Deodorant and antibacterial soaps are also growing in popularity although care has to be taken not to include skin sensitizing agents or components which interact with the other soap ingredients. Hexachlorophene and

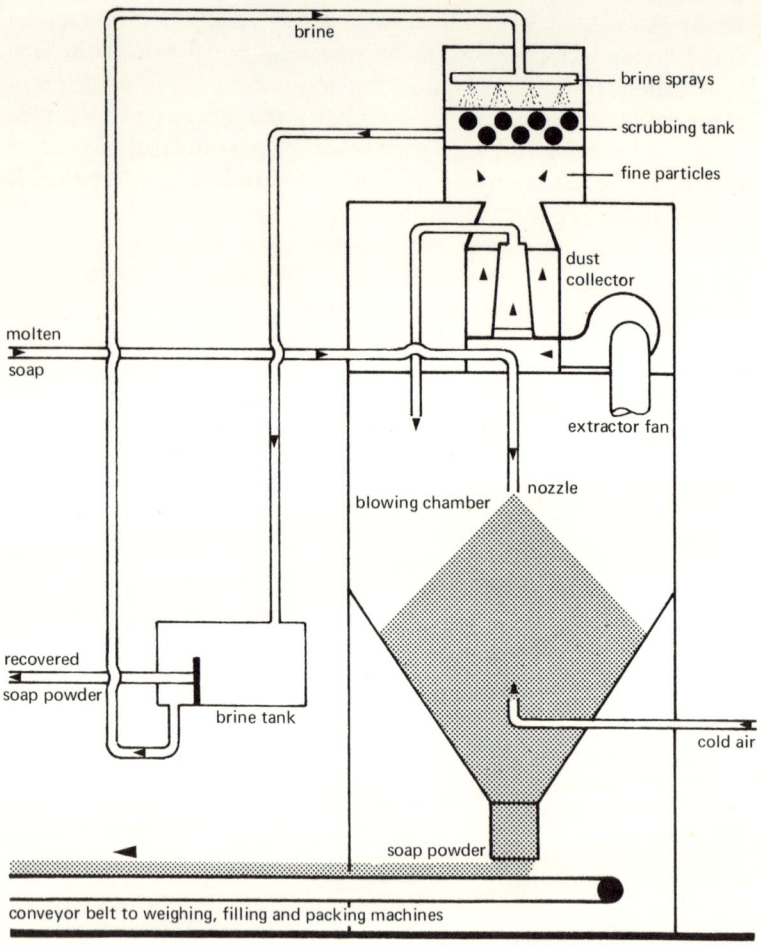

Fig. 3.11   Manufacture of soap powder

3,4,4-trichlorocarbanilide (TCC) have been used successfully, although phenolic substances tend to discolour in sunlight and soaps containing these ingredients are usually deeply coloured to mask this effect.

Soap flakes are produced by spraying molten soap on to a chilling roll from which it is removed in the form of small chips. After drying, the chips are rolled between steel rollers in a pre-mill and passed through a series of reducing rollers until the soap forms a very thin skin. This is then cut into diamond shapes by a rotating knife.

Soap powders are produced by mixing molten soap with suitable builders and drying by spraying through a nozzle countercurrent to a blast of cold air. The soap dries as small granules which drop on to a conveyor belt which takes them to packaging machines. The air from the drying tower is scrubbed with a brine spray to remove fine soap dust which is later recovered.

Transparent soap can be prepared by dissolving neat soap in ethanol, concentrating the solution by evaporating off most of the alcohol and then running into moulds. A more usual method is to use a blend of fats containing a high proportion of castor oil, and prepare the soap by the cold process in order to retain the glycerol formed.

Readily soluble soaps with good lathering properties, prepared from blends of coconut oil and palm oil, can be used for washing with sea water.

Scouring powders are made by mixing 2–5% of a detergent base with water softening builders and a substantial quantity of a mild abrasive such as pumice, silica or feldspar. Recently bleaches such as chlorinated trisodium phosphate have been added.

Liquid soaps for washing-up and shampoos are aqueous solutions of soft soaps made from coconut oil and olive oil blends, but these have now been almost completely replaced by soapless detergent liquids.

Floating soap is said to have been discovered accidentally in 1879 when bubbles of air became trapped in a batch of soap which had been excessively stirred during manufacture. Soaps of this kind are not milled and have a high moisture content.

The discarded lye from soap-making processes contains between 5 and 10% of glycerol, and this valuable by-product is recovered by a vacuum distillation process.

After standing, any soap which separates out is filtered off and the filtrate acidified and treated with ferric chloride. The solution is then re-filtered to remove the precipitated insoluble iron soaps, neutralized with lime and again filtered to remove excess iron as ferric hydroxide. Multistage evaporators remove most of the water, any salt which separates being centrifuged off. The resulting first stage glycerol is about 80–85% pure and this is further refined by low pressure steam distillation and treatment with charcoal.

Commercial grade glycerol is about 98% pure and is used in the manufacture of explosives, alkyd resins, dyestuffs and fat-extenders. Glycerol is also used as a plasticizer in the manufacture of cellophane, and as a moisturizing agent ('humectant') in the manufacture of printing and duplicating inks. A re-distilled product of high purity is used pharmaceutically, in cosmetics, and as a food preservative in cakes and confectionery.

SOAPLESS DETERGENTS

Although an excellent detergent, soap suffers from the serious drawback of producing sticky insoluble scums with solutions of metal salts. Thus laundering in hard water causes soap wastage and greying of fabrics, while the presence of soap scum leads to serious difficulties in dyeing processes. In addition dyestuffs are often sensitive to the alkaline conditions produced by the hydrolysis of soap in aqueous solution, while soap decomposes in the acid solutions required for dyeing and other textile processes.

Since 1930 a number of synthetic detergents have been produced which do not contain soap and are free from the disadvantages mentioned. It is convenient to discuss these according to their ionization characteristics.

ANIONICS

The anionics, where the ionized 'tail' unit carries a negative charge, comprise the largest group of detergents and account for about 95% of all those in common use. The three main classes of anionics are alkyl sulphates, alkyl aryl sulphonates and the soaps.

The alkyl sulphates were the first type of synthetic detergent to be produced and until about 1948 were the most commonly used ingredients of soapless detergent powders and liquids. The first of the commercially important alkyl sulphates were the sulphated oils discovered by Frémy in 1831. Sulphated castor oil ('Turkey Red Oil') is still in use today as a dyeing assistant, an emulsifying agent for cut-

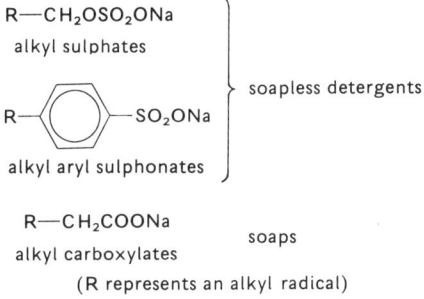

ting oil and agricultural sprays and as a leather dressing. It is the only sulphated oil of importance, although sulphated fish oils are also produced commercially.

Turkey Red Oil is prepared by stirring concentrated sulphuric acid into castor oil maintained at a temperature of 25 °–30 °C. After the reaction is complete, water is added to float off the upper layer of sulphated oil which is then partially neutralized with sodium hydroxide solution. The process of sulphation mainly involves the hydroxy groups of the ricinoleic acid present in the castor oil.

$$CH_3(CH_2)_6\underset{\underset{\text{ricinoleic acid}}{OH}}{C}HCH_2CH{=}CH(CH_2)_6COOH \longrightarrow$$

$$CH_3(CH_2)_6\underset{\underset{\text{sulphated ricinoleic acid}}{OSO_2OH}}{C}HCH_2CH{=}CH(CH_2)_6COOH$$

Alkyl sulphates prepared from the higher alcohols such as lauryl and cetyl alcohol were the first widely used synthetic detergents. They are synthesized by the catalytic hydrogenation of the appropriate fatty acid ester. The latter is prepared either by the esterification of the fatty acids produced by splitting suitable fats such as tallow and coconut oil, or by the heating of fats under pressure with methanol (transesterification).

Hydrogenation of the esters is carried out at elevated temperature ($\rightarrow$300 °C) and high pressure (30 MN/m^2 (300 atmospheres)) in the presence of a copper chromite catalyst. Since any unsaturated bonds are also reduced under these conditions the fatty alcohols produced in this way are always saturated.

$$R-\overset{\overset{\text{O}}{\|}}{\text{C}}OCH_3 \xrightarrow{\text{H}_2\ +\ \text{copper chromite}} R-CH_2OH + CH_3OH$$

fatty acid　　　　　　　　　　　　　　　　fatty alcohol　methanol
methyl ester

Cetyl and oleyl alcohols can also be obtained directly by distillation of sperm oil. The alcohols are sulphated by treatment with chlorosulphonic acid or sulphuric acid, the reaction mixture being cooled with ice-water and neutralized with sodium hydroxide.

$$R-CH_2OH \qquad R-CH_2OSO_2OH \qquad R-CH_2OSO_2ONa$$

primary (fatty) alcohol　　　　sulphated alcohol　　　　sodium alkyl sulphate

The most important of the alkyl sulphates is sodium lauryl sulphate, produced from coconut and palm oils. It is a good foaming agent and emulsifier, and owing to its mildness is widely used in shampoos, cosmetics and toothpastes. The cetyl, oleyl and stearyl derivatives are also good detergents and are used for scouring textiles and in the tanning and dyestuff industries. As with the other syndets the calcium and magnesium salts of these detergents are water soluble and they are therefore relatively unaffected by hard water.

Sulphated amides are also valuable detergents with particularly good foaming properties. These are usually prepared by condensing a fatty acid with an ethanolamine and sulphating the product.

$$R-COOH + \overset{\overset{\text{NH}_2}{|}}{\text{C}}H_2CH_2OH \longrightarrow$$

fatty acid　　monoethanolamine

$$R-CONHCH_2CH_2OH \xrightarrow{\text{H}_2\text{SO}_4} R-CONHCH_2CH_2OSO_2OH$$

ethanolamide　　　　　　　　　　　sulphated amide

Secondary alkyl sulphates made from petroleum base materials have been manufactured by the Shell Petroleum Co. since 1942, using a process developed in the Royal Dutch/Shell Group. The commercially important 'Teepol' is produced from a fraction containing olefins with straight chains of from 8 to 18 carbon atoms, obtained by thermally cracking paraffin wax. Sulphation of this fraction is carried out using concentrated sulphuric acid under carefully controlled conditions, producing mainly monoalkyl hydrogen sulphates with some dialkyl derivatives.

$$R-CH=CH_2 \xrightarrow{H_2SO_4} R-CH\begin{smallmatrix}CH_3\\\\OSO_2OH\end{smallmatrix} \xrightarrow{NaOH} R-CH\begin{smallmatrix}CH_3\\\\OSO_2ONa\end{smallmatrix}$$

| olefin | secondary alkyl sulphate | sodium secondary alkyl sulphate |

(R is a hydrocarbon chain of $C_{10} - C_{18}$)

The reaction mixture is neutralized with sodium hydroxide solution and treated to remove the dialkyl derivatives and unsulphated hydrocarbons. The resulting product contains about 20% sodium alkyl sulphates but this can be increased if required by evaporation under reduced pressure.

'Teepol' is supplied as a liquid, spray dried powder or in paste form. It has excellent wetting and detergent properties and is widely used in the preparation of domestic cleaning materials such as carpet shampoos, and industrially for degreasing and scouring, wetting textiles, stabilizing paint emulsions, dispersing pigments and general cleaning purposes.

The disadvantages of scum formation and acid decomposition encountered with soaps were obviated in the design of the alkyl sulphate type detergents by replacing the vulnerable carboxylate unit with the sulphate group. During World War II the Germans produced primary and secondary alkyl sulphonates, the sulphonate radical playing a similar role to that of its sulphate counterpart, except that the sulphur atom was now directly linked to the first chain carbon, without a bridging oxygen atom.

$$R-CH_2OSO_2ONa \qquad\qquad R-CH_2SO_2ONa$$

sodium alkyl sulphate	sodium alkyl sulphonate

Production of the alkyl sulphonates was originally carried out by the Reed reaction, named after its American discoverer. Saturated

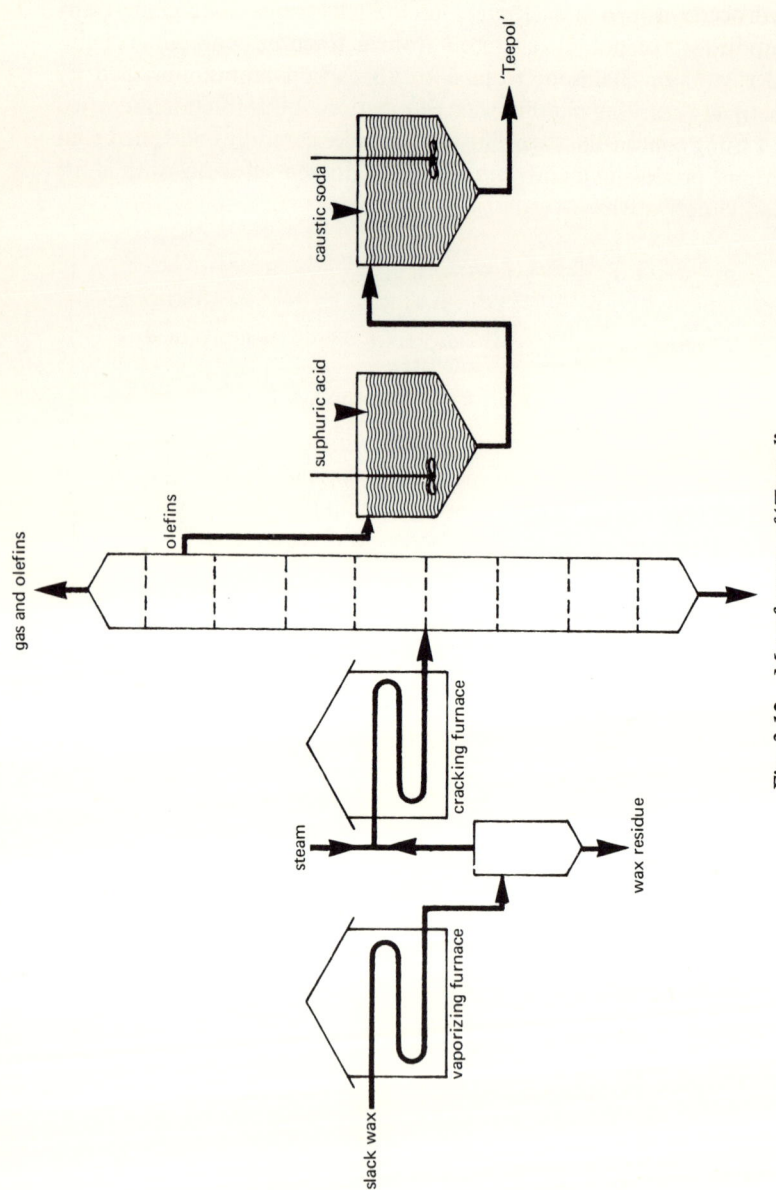

Fig. 3.12 Manufacture of 'Teepol'

hydrocarbon mixtures obtained from the Fischer-Tropsch process (production of straight chain paraffins and olefins from mixtures of hydrogen and carbon monoxide using a cobalt/thorium oxide catalyst) were treated at low temperature (30 °–40 °C) with a mixture of chlorine and sulphur dioxide in the presence of ultraviolet light. The sulphonyl chlorides formed were hydrolysed with sodium hydroxide to give the sodium alkyl sulphonates.

$$R—CH_3 \xrightarrow[\text{ultraviolet light}]{SO_2 + Cl_2} R—CH_2SO_2Cl \xrightarrow{NaOH} R—CH_2SO_2ONa$$

| paraffin | alkyl sulphonyl chloride | sodium alkyl sulphonate |

(R represents a hydrocarbon chain of $C_{14} \longrightarrow C_{16}$)

A modified one-stage variant of the Reed process known as the Hostapon process involves the treatment of paraffins with a mixture of oxygen and sulphur dioxide.

In 1930 the German firm of I.G. Farbenindustrie marketed what were known as 'blocked soaps' under the trade name of 'Igepons'. They overcame the trouble of the vulnerable carboxylate unit of the soap molecule by blocking it with a short chain compound having an active group at either end. This permitted one active group to react with the carboxylate group of the soap while the other provided the hydrophilic head unit of the detergent. The most successful of the 'Igepons' was 'Igepon T' in which the blocking compound was methyl taurine. 'Igepon T' is an outstanding detergent and wetting agent and is used extensively in the textile industry.

$$C_{17}H_{33}\overset{\overset{O}{\|}}{C}Cl + \left[\overset{\overset{CH_3}{|}}{N}HCH_2CH_2SO_2O\right]^{-}Na^{+} \xrightarrow{NaOH} \left[C_{17}H_{33}\overset{\overset{O}{\|}}{C}—\overset{\overset{CH_3}{|}}{N}CH_2CH_2SO_2O\right]^{-}Na^{+}$$

| oleyl chloride | sodium salt of methyl taurine | 'Igepon T' |

The successful blocking of the carboxylate group led to the development of another sulphonate type detergent in which the hydrocarbon chain contains a benzene ring. Sulphonation of the benzene ring is much easier than sulphonation of a straight paraffin chain. The benzene must still be alkylated, however (i.e. have a chain hydrocarbon linked to it), to preserve the hydrophobic/hydrophilic balance of the detergent molecule. This is usually carried out by stirring benzene with propylene tetramer (i.e. a polymer of propylene containing four monomer units) to give what is termed 'alkylate'. After sulphonating the alkylate with concentrated sulphuric acid at a temperature of 40 °–45 °C, water is added and after settling, the lower acid layer is run off. The alkyl aryl sulphonate is then neutralized with sodium hydroxide solution and spray dried.

$$H_3C \diagdown \diagup (CH_2)_9CH_3$$
$$CH$$

benzene + $CH_2{=}CH(CH_2)_9CH_3$ $\xrightarrow{\text{HF catalyst}}$ dodecyl benzene (alkylate) $\xrightarrow[< 40\,°C]{H_2SO_4}$

benzene   dodecylene (propylene tetramer)   dodecyl benzene (alkylate)

$$H_3C \diagdown \diagup (CH_2)_9CH_3$$
$$CH$$

$$SO_2OH$$

dodecyl benzene sulphonic acid

Alkyl aryl sulphonates of this type, widely used in Europe and the USA as components of washing powders and wetting and dispersing agents, were first marketed in 1933 by I.G. Farbenindustrie as 'Igepol NA' and three years later in the USA as 'Nacconol NR'. Although the alkyl aryl sulphonates are attractive because of the ready availability of propylene they have one serious disadvantage. They are resistant to biological breakdown because they form stable complexes with the proteinaceous material found in watery waste. This produces uncontrollable quantities of foam on rivers containing effluent and on the settling ponds in sewage farms. Detergents of this type are referred to as 'hard' or 'non-biodegradable' and are now little used in highly industrialized countries. A number of 'soft' linear alkyl aryl sulphonates are now available which are based upon straight chain paraffins or olefins, together with primary alkyl sulphates derived from 'Oxo' alcohols. Biodegradable detergents are now the only type permitted by law in certain countries (e.g. West Germany) although the propylene tetramer type of alkyl aryl sulphonate is still preferred in less highly populated countries having no effluent problems.

CATIONICS

In cationic detergents the long chain hydrophobic tail unit forms the cation, i.e. carries a positive charge on ionization. Many textile fibres (and other surfaces) are negatively charged and such cations tend to attach themselves to these surfaces. This makes cationics good softening agents, although as cleansing agents they are inferior in performance.

The only cationics which are of major importance commercially are the quaternary ammonium compounds which can be regarded as substituted ammonium halides. These are stable and readily soluble and were first prepared in the 1890's, although it was not until 1908 that their bactericidal properties were recognized. Domagk, who was awarded the Nobel Prize for his work in developing the sulphonamides, described a number of powerful bactericides of this type in 1935 and commercial production began just before World War II.

The alkyl methyl ammonium halides are prepared by the reaction of aliphatic amines with alkyl halides in the presence of sufficient sodium hydroxide to neutralize the hydrochloric acid produced. The mono-, di- and tri-alkyl derivatives are prepared using primary, secondary and tertiary amines respectively.

$$\text{R—NH}_2 \ + \ 3\text{CH}_3\text{Cl} \ \xrightarrow{\ 2\text{NaOH (aq.)}\ } \ \left[\text{R—N}^{\underset{\underset{\text{CH}_3}{|}}{\overset{\overset{\text{CH}_3}{|}}{}}\text{—CH}_3 \right]^+ \text{Cl}^- \ + \ 2\text{NaCl} + 2\text{H}_2\text{O}$$

primary   methyl
aliphatic   chloride
amine   alkyl trimethyl
                     ammonium chloride

$$\text{R—NH—R} \ + \ 2\text{CH}_3\text{Cl} \ \xrightarrow{\ \text{NaOH (aq.)}\ } \ \left[\text{R—N}^{\underset{\underset{\text{CH}_3}{|}}{\overset{\overset{\text{CH}_3}{|}}{}}\text{—R} \right]^+ \text{Cl}^- \ + \ \text{NaCl} + \text{H}_2\text{O}$$

secondary   methyl
aliphatic   chloride
amine   dialkyl dimethyl
                   ammonium chloride

(R represents an alkyl radical of $C_8 \longrightarrow C_{18}$)

7.

The alkyl trimethyl ammonium halides are employed as textile softeners, bactericides and in antistatic preparations for plastic articles. Cetyl trimethyl ammonium bromide is widely used as a skin antiseptic in creams, ointments and cosmetics, and in pastilles for ulcerated mouth conditions. Mixtures of quaternary ammonium bactericides of this type with high performance detergents such as 'Teepol' are available commercially for washing dishes, bottles and cutlery in canteens and factories, e.g. 'Nonidet J'. The dialkyl di-methyl derivatives are used as corrosion and mould inhibitors and for emulsifying cutting oils.

Alkyl pyridinium chlorides and alkyl dimethyl benzyl ammonium chlorides are also widely used as bactericides, mould retardants, dyeing assistants, and for pharmaceutical purposes.

$$\left[C_{16}H_{33}-\overset{\overset{\displaystyle CH_3}{|}}{\underset{\underset{\displaystyle CH_3}{|}}{N}}-CH_3 \right]^{+}Br^{-} \qquad \left[\bigcirc\!\!\!\!\!\bigcirc N-CH_2(CH_2)_{14}CH_3 \right]^{+}Cl^{-}$$

'Cetrimide' (ICI)   cetylpyridinium chloride
(bactericidal creams)   (antiseptic pastilles)

NON-IONICS

The non-ionic surfactants do not ionize in solution and therefore cannot react with the metallic salts in hard water to form scums. Since they are also stable to acids and alkalis they are ideal for use in the textile processing and dyeing industries. The hydrophobic part of the surfactant molecule is solubilized by the introduction of groups

such as $\diagdown O \diagup$, —OH and $-C\!\!\begin{array}{c}\diagup O \\ \diagdown O^-\end{array}$. Some of the foam-producing

glucosides which occur in nature, such as the saponin present in horse chestnuts, are examples of non-ionizing surfactants.

The non-ionics are generally liquids or pastes and often have poor detergent properties but are useful as emulsifying agents, dyeing assistants and for enhancing the activity of other detergents (syner-gistic effect). The most important non-ionic detergents are prepared by the reaction of long chain alcohols (or phenols), fatty acids or fatty amines, with ethylene oxide. This produces polymeric ethoxy chains containing the ether linkage $\diagdown_O\diagup$ and terminating in a hydroxyl group —OH. The solubility and surface activity of these compounds is dependent on the length of the ethoxy chain.

$$R-\!\!\left\langle\bigcirc\right\rangle\!\!-OH + n\left[\begin{array}{c}H_2C\\ |\\ H_2C\end{array}\!\!>\!\!O\right] \xrightarrow[140-180\,°C]{NaOH\ (aq.)} \left\langle\bigcirc\right\rangle\!\!-\!\!\left[(CH_2)_2-O\right]_n\!\!H$$

alkyl phenol   ethylene oxide   alkyl phenol/ethylene oxide condensate

(**R** represents an alkyl radical of $C_8 \longrightarrow C_{10}$; *n* is an integer from 6−12)

Alkyl phenol/ethylene oxide condensates of the type shown are viscous liquids with good detergent properties which are widely used in the scouring and laundering of textiles ('Lissapol N'—ICI). Condensates of ethylene oxide chains with fatty alcohols, fatty amines or fatty acids are also liquids which are used as dyeing assistants, emulsifying agents and scourers.

$$R-COOH + n\left[\begin{array}{c}H_2C\\ |\\ H_2C\end{array}\!\!>\!\!O\right] \xrightarrow{140-180\,°C} R-COO\!\!\left[(CH_2)_2-O\right]_n\!\!H$$

long chain
fatty acid

(**R** represents an alkyl radical of $C_{10} \longrightarrow C_{12}$; *n* is an integer from 6−12)

Alkylolamines such as ethanolamine are sometimes used, instead of ethylene oxide, to form condensates with long chain fatty acids (C_{10}–C_{12}) such as lauric acid. The resulting surfactants have good foaming properties and improve other detergents (synergistic effect).

$$\underset{\text{lauric acid}}{C_{12}H_{25}COOH} + \underset{\text{monoethanolamine}}{\overset{\overset{\textstyle NH_2}{|}}{CH_2CH_2OH}} \xrightarrow{-H_2O} \underset{\text{lauryl ethanolamide}}{C_{12}H_{25}CONHCH_2CH_2OH}$$

Also of importance are the monoesters of fatty acids with polyhydric alcohols (polyols) such as glycerol and pentaerythritol. These are without detergent properties but are very valuable emulsifiers and are used in the preparation of cosmetics, liquid polishes, ice-cream, margarine, synthetic cream and as a fat extender in cooking fats.

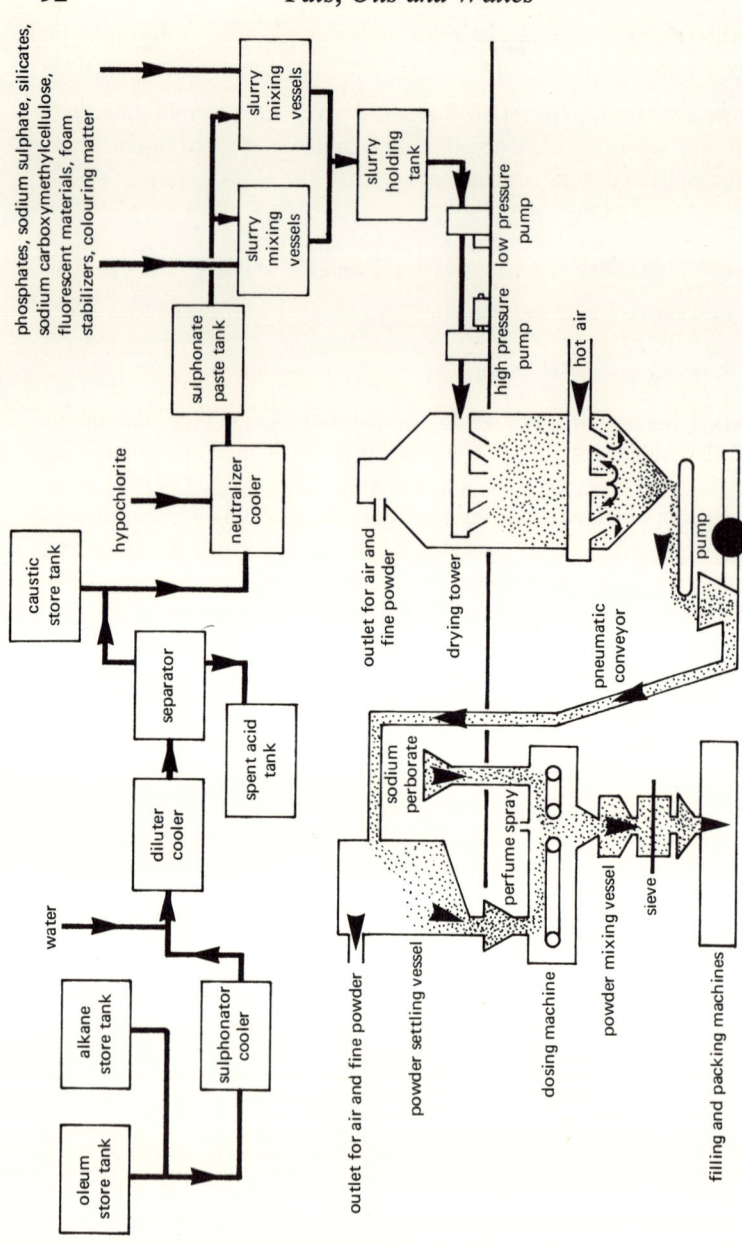

phosphates, sodium sulphate, silicates, sodium carboxymethylcellulose, fluorescent materials, foam stabilizers, colouring matter

slurry mixing vessels

slurry mixing vessels

slurry holding tank

low pressure pump

high pressure pump

hot air

sulphonate paste tank

neutralizer cooler

hypochlorite

caustic store tank

separator

spent acid tank

diluter cooler

water

alkane store tank

oleum store tank

sulphonator cooler

outlet for air and fine powder

drying tower

pump

pneumatic conveyor

sodium perborate

powder settling vessel

perfume spray

dosing machine

powder mixing vessel

sieve

outlet for air and fine powder

filling and packing machines

Fig. 3.13   Manufacture of soapless detergent powder

Solid non-ionic detergents can be produced by attaching a poly-propylene oxide chain to the polyethylene oxide chain. The former must have a molecular weight of at least 1000 and the polyethylene oxide content must be in the region of 70%. Detergents of this type are marketed in the USA by the Wyendette Chemicals Corporation ('Pluronics').

Manufacture of Soapless Detergent Products

Soapless detergent powders are manufactured by spraying a slurry of the ingredients down a drying tower, countercurrent to a blast of hot air. The perfume and sodium perborate are added after spray drying because of their sensitivity to heat. The dried powder is continuously removed from the base of the tower by a belt conveyor. From here it is blown pneumatically up to an overhead hopper from which it passes with metered quantities of sodium perborate under a perfume spray to the packaging machines. Any fine dust is separated from the powder by sieving and returned to the slurry plant for retreatment.

Liquid soapless detergents used for dish-washing are a mixture of a sulphonate with a small proportion of a very soluble non-ionic detergent and a foaming agent. Bleach, brighteners and alkaline builders are omitted.

Top of tower used for spray drying soapless detergent

Special Applications of Surfactants

In addition to their applications in laundering and domestic cleaning, surfactants are widely used as wetting and dispersing agents, lubricants and emulsifiers. Thus metal soaps such as zinc stearate are used as high temperature lubricants, mould dressings and leather softening oils and in compounding greases for slipways, heavy machinery and rust protection. Surfactants also act as emulsifying agents in the cosmetics and food industries and in the formulation of agricultural sprays. Paint pigments are dispersed effectively by surfactants to give better brushing and covering characteristics. In the dyeing industry surfactants are of great value in promoting even hues and producing a satisfactory suspension in the case of disperse dyes.

POLISHES

The modern wax-based polishes were first introduced as wax solvent pastes in the late nineteenth century. Up till then floors, furniture and other surfaces were treated by a variety of methods such as scrubbing, oiling, sanding, varnishing and French polishing. Waxes such as beeswax had been used long before this for treating wooden surfaces but these had to be laboriously applied by rubbing with a block of the material concerned. Later, pastes of beeswax in turpentine or emulsions of wax in soda solutions were used, but these again required the expenditure of vast amounts of time and energy to achieve a satisfactory surface gloss.

Several factors were probably responsible for the genesis of the modern labour-saving paste and emulsion polishes at the beginning of this century. The close of the Victorian era saw fewer women entering domestic service and less cheap labour in the home. This, together with the increasing emancipation of women in the twentieth century, led to demands for more labour-saving materials. At the same time the rise of organic chemistry and the development of the petroleum industry provided new ingredients for the formulation of commercial polishes giving better results with far less labour.

Even today new manufacturing techniques and the inclusion of additives such as silicone are continually being experimented with to improve the finished product. Another important task often con-

Testing a paste polish
for firmness

fronting the manufacturer is to use poorer quality or cheaper alternative raw materials and still produce a high quality polish.

Although a number of special wax polishes have been developed for application to specific surfaces such as footwear, motor cars, furniture and floors, two principal types of polish can be distinguished. There are those in which the wax base is dissolved in a nonaqueous solvent (paste polishes) and those in which the wax is in the form of an aqueous emulsion (liquid polishes). The type of polish determines the method of manufacture. In both cases the waxes to be used are broken up in crushing machines ('impact mills') and then melted in steam- or electrically-heated pans. Mixing is usually carried out with electrically operated propeller type stirrers, although in smaller works stirring is still done by hand with a wooden paddle or iron spatula. The molten wax blend is run into a mixing vat through heated pipes where it is either stirred into an appropriate solvent (paste polishes), or rapidly stirred into water and then emulsified in a colloid mill (liquid polishes). The polish is put into tins, bottles or aerosols in the liquid state using vacuum operated filling machines.

NATURAL AND SYNTHETIC WAXES

Initially beeswax was used almost exclusively for polishes but the modern polish manufacturer has an amazing variety of natural and synthetic waxes to choose from. It is easier to classify these according to their origin rather than their chemical constitution as they are usually complex mixtures of wax alcohols and wax acid esters.

(a) The vegetable waxes are important polish ingredients as they are usually hard and thus impart a high gloss and form a highly protective surface. The most well known is *carnauba wax*, which is scraped from the leaves of the carnauba palm, a native of Brazil. After beating the sun dried leaves with sticks, the dislodged crude wax is melted over an open fire with a little water, and poured into moulds to give a grey impure commercial product. Annual production is of the order of 12 000 tonnes—the yield from each tree being about 170 g (6 oz) a year. The pure wax is almost colourless, very hard and has the highest melting point of all the vegetable waxes (82 °–90 °C). On cooling the molten wax, or a mixture containing carnauba wax, characteristic rings appear due to the crystallizing out of myricyl alcohol (C_{29}). This is an important sales feature of the manufacture of shoe polishes. The wax also contains ceryl alcohol (C_{27}), melissyl alcohol (C_{30}) and the myricyl ester of cerotic acid (C_{26}), together with a mixture of wax fatty acids and hydrocarbons.

Other tree waxes of similar properties are *raffia wax*, obtained from the leaves of the Madagascan sago palm, *ouricuri wax* and *cauassú wax* obtained from species of Brazilian palm, and *palm wax* which is collected from the trunk of the South American wax palm.

Candelilla wax, which is often used as a substitute for carnauba wax, is obtained from the stems of a weed which grows prolifically in Mexico. The cut stems are boiled in water and the crude wax skimmed off the surface as a brownish, aromatic smelling liquid which is poured into moulds to solidify, and then broken up. Wax is also extracted from *esparto grass* in North Africa using solvent extraction techniques (benzene) which are also used in the production of *cotton wax*—the latter being unusually soft for a vegetable wax and resembling beeswax.

Recently plant has been set up for the recovery and refining of *sugar cane wax*. During the crushing of sugar cane, part of the wax is carried over into the syrup and this is filtered off together with the chalk and other solids during the carbonation stage of refining. The wax is usually recovered from the press cake by solvent extraction (acetone), and is a hard greenish solid. Bleaching can then be carried out to give a virtually colourless product of complex chemical composition containing about 60% wax esters, 25% fatty acids and 5% wax alcohols. In the Hatt process the wax is extracted from the press cake by demineralization with hydrochloric acid, followed by vacuum distillation and bleaching with a mixture of sulphuric and chromic acids. Although at present extraction of sugar cane wax is only carried out on a small scale in Cuba, Australia and the USA, the potentialities of this source are very great because of the enormous quantities of sugar which are refined annually.

(b) The most important of the animal waxes is *beeswax* and it has been used for many hundreds of years as a component of such preparations as cosmetics, polishes and sealing compounds. It is produced by the wax glands in the abdominal wall of the honeybee in the form of thin scales. These scales are fabricated into the wax comb which is then filled with honey. After the honey has been removed by draining and centrifuging, the wax is melted by heating in water and filtered. Natural beeswax is a pale yellow colour and can be bleached by exposure to the sun or by using oxidizing agents. It has a characteristic smell and a faint taste, and is sticky and easily moulded at body temperature.

About 70% of beeswax consists of wax esters, mainly myricyl palmitate and other myricyl esters. There is also an unusually high hydrocarbon content (12%) including nonacosane (C_{29}) and hentriacontane (C_{31}). The stickiness of beeswax causes drag when applying it to surfaces and this has to be counteracted by admixture with other waxes or fatty acid esters. Owing to its complex nature it is difficult to find solvents which dissolve beeswax completely, aromatic hydrocarbons such as benzene being most successful.

A special type of beeswax known as *Ghedda wax* is imported from the Far East and is sometimes used to adulterate beeswax. Another

	m.p. (°C)	Colour	Origin	Notes
(a) Vegetable waxes				
Carnauba	85–90	grey-yellow	carnauba palm, Brazil	hard, dense, brittle
Cauassú	82–84	yellow-brown	cauassú plant, Amazon	similar to carnauba
Candelilla	68–72	green-brown	Mexican weed	semi-hard
Ouricuri	83–85	grey-yellow	ouricuri palm, Brazil	similar to carnauba
Raffia	79–83	brown	sago palm, Madagascar	hard
Palm	83–85	pale brown	wax palm, S. America	similar to carnauba
Esparto	73–74	brown	esparto grass	hard, non-crystalline
Sugar cane	76–82	grey-white	sugar cane stem	sticky, semi-hard
Cotton	68–71	green-brown	cotton fibres	resembles beeswax
(b) Animal waxes				
Beeswax	62–72	yellow-white	honeybee	kneads, sticky
Ghedda	63–71	brown-yellow	bee, Far East	similar to beeswax
Chinese insect	80–83	yellow-white	privet insect, China	very hard, translucent
Shellac	75–80	red-brown	lac insect, India	hard, glossy
Spermaceti	42–46	white	whale	silky, translucent, brittle
Lanolin	32–41	yellow	wool	soft grease
(c) Mineral waxes				
Paraffin	low m.p. 40–42 / high m.p. 65–70	white	petroleum	crystalline, translucent
Ozokerite	63–86	yellow	fossil wax, USSR, Romania	microcrystalline, high m.p.
Montan	60–76	brown-yellow	brown coal	hard, brittle, very dense
Syncera	68–70	yellow	petroleum	microcrystalline

Naturally occurring waxes used in polishes

wax imported from China, called *Chinese insect wax*, is produced by a small insect which feeds on a variety of privet native to that country. *Shellac wax* is also of insect origin, occurring in shellac resin to the extent of about 4 to 6%. When shellac is dissolved in alcohol to produce French polish the wax remains undissolved and can be separated off. It is unusually hard and produces surface films with a high gloss.

Other animal waxes include *Spermaceti* which is extracted from sperm whale oil, and *lanolin* ('wool-fat') which is present in raw sheep's wool. Both form stable emulsions and thus find some application in the making of liquid polishes. The structure of lanolin is interesting as it contains esters of the higher fatty acids with complex cyclic alcohols such as cholesterol (C_{27}) and lanosterol (C_{30}).

(c) The mineral or hydrocarbon waxes are associated with the fossil fuels oil and coal. *Paraffin wax* is composed of straight chain paraffin hydrocarbons together with some branched and aromatic hydrocarbons, the melting point rising with the molecular weight of the constituents. 'Distillate' wax is produced by chilling wax-containing feedstocks and filtering. The lower melting point waxes are removed by heating to a temperature just high enough to liquefy them (sweating) and then pressing to remove the trapped oil. A high quality wax with a very fine crystalline grain is obtained by the extraction of lubricating oil with a solvent mixture such as methyl ethyl ketone and benzene. The 'slack wax' so produced is then further refined by distillation, in order to separate it into narrow melting point fractions of high purity. The product is termed microcrystalline wax and is particularly suitable for polish formulations. The paraffin waxes are colourless, odourless and tasteless with a translucent appearance and a slightly sticky feel. They are soluble in most of the usual organic solvents except alcohols.

Natural deposits of paraffin wax are found within fissures in the rocky strata around certain oilfields, especially in eastern Europe, Asia and Texas. It is clear that seepage of crude oil along the fissures has led to evaporation of the lower boiling point fractions followed by pressure filtration through porous earth, leaving a dark coloured waxy residue. This is known as *'ozokerite'* or 'earth wax' and is a valuable form of paraffin wax, resembling the microcrystalline variety in having branched paraffin chains, which partly explains its high melting point. Ozokerite wax can be bleached to a yellow or white, but is often used in shoe polishes in its natural dark form, and is an important substitute for carnauba wax.

Another dark coloured fossil wax called *montan wax*, which is widely used for shoe polishes, is extracted from brown coal. The coal is first crushed and then extracted with a solvent such as benzene, the product being further refined by steam distillation at reduced pressure, and bleaching. Montan wax is unusually complex in make-up, containing a number of free fatty acids, alcohols, ketones and resins, together with wax esters. It is only readily soluble in aromatic and chlorinated hydrocarbons. Because of its hardness, high melting point and ease of mixing with other waxes, montan also makes a suitable substitute for the more expensive carnauba wax, and attempts have been made to produce it from peat and lignite deposits in the UK.

In addition to the natural waxes mentioned above, a number of synthetic or modified waxes have been produced. The *Carbowaxes* are based on polyethylene glycol which is produced when ethylene oxide is passed under pressure into ethylene glycol at 130 °–150 °C with a little sodium hydroxide as catalyst.

$$n\begin{bmatrix}H_2C\\ \ |\ \ >O\\ H_2C\end{bmatrix} + 2\ \begin{matrix}CH_2CH_2\\ |\ \ \ |\\ OH\ OH\end{matrix} \xrightarrow[130-150\ °C]{NaOH\ (aq.)} \begin{matrix}CH_2CH_2\\ |\ \ \ |\\ OH\ O\end{matrix}\begin{bmatrix}-(CH_2)_2-O\end{bmatrix}_n\begin{matrix}OH\\ |\\ CH_2CH_2\end{matrix}$$

ethylene   ethylene   polyethylene glycol
oxide   glycol

(*n* is an integer from 6–12)

The Carbowaxes, with a molecular weight from 200 to 700, are liquids which may be esterified to form useful wax emulsifiers, in addition to themselves acting as waxes. The most commonly used emulsifying agent for water-containing polishes is polyethylene glycol stearate. Carbowaxes with even higher molecular weights from 1000 to 4000 are water soluble solids which also dissolve readily in most organic solvents, but only sparingly in aliphatic hydrocarbons. These are also used as emulsifying agents for polishes and have useful mould suppressing properties.

Blends of the higher aliphatic alcohols synthesized from the corresponding fatty acids by catalytic hydrogenation are also employed as easily emulsified wax substitutes in the manufacture of polishes, soaps and leather dressing. The most well known example of this type is a blend of palmityl and stearyl alcohols known as Lanette wax.

$$CH_3(CH_2)_{16}COOH + 2H_2 \xrightarrow[\text{30 MN/m}^2 \text{ (300 atm.)}]{\text{copper chromite}} CH_3(CH_2)_{16}CH_2OH + H_2O$$

stearic acid   stearyl alcohol

Recently a new class of synthetic wax has appeared which has been termed a 'pseudo-ester' wax, since it differs from the true ester waxes in being produced from fatty acids and amines instead of fatty acids and alcohols. The best known are the Abril waxes which are available in a wide range of different types. Lower melting point Abril waxes are used in cosmetics and polishes as a cheaper substitute for beeswax. The higher melting point varieties (140 °–160 °C) are used as corrosion inhibitors and plasticizers.

Higher fatty amines known as Armids and Armowaxes (Chemical Division of Armour and Co. Ltd) have also recently become commercially available. They range from soft to hard waxy solids which are insoluble in water but fairly soluble in many hot organic solvents. They are produced by the dehydration of fatty acid ammonium salts in the presence of ammonia gas and are used as paint thickeners, detergent builders, mould release agents and dispersants in the rubber industry.

Modified waxes include the chlorinated paraffin waxes such as 'Chlorax' (Watford Chemical Co.) which are used for polishes, waterproofing and in linoleum manufacture, and the German I.G. waxes which are produced by treating montan wax. The wax is first deresinated by treatment with ethanol and then oxidized or esterified.

WAX POLISHES

These include polishes for floors, furniture, motor cars and shoes. They are mixtures of waxes and solvents which may or may not include water, together with special additives such as resins, silicone oils, pigments, abrasives, antiseptics and emulsifying agents to improve the final product. Their function is to clean the surface to which they are applied and to cover it with a smooth, clear film, which has a high gloss and protects the surface from spilled liquids, scuffing, staining and other normal wear and tear.

The use of solvents and aqueous emulsions of wax allows the polish to be applied easily. After evaporation of the solvent or water, this leaves a smooth film of wax which can be further glossed by rubbing or brushing. The original commercial beeswax polishes which were used in Victorian days were general purpose polishes which could serve a number of different purposes. Gradually, however, polishes began to appear which had been specially formulated for treating the new types of flooring and wood and metal finishes which were being developed. Although 'all-purpose' polishes are still available there are certain limits to their use to prevent surface damage, floor slip or discoloration.

Furniture polishes are basically wax blends which are either mixed into a solvent such as turpentine, white spirit or a drying oil to produce a paste, or emulsified with water to give creams which can be applied with a cloth, or using an aerosol spray. The furniture creams have the advantage that, containing both water and an oily solvent, they are able to remove both oil- and water-borne stains. Also, since they contain an emulsifying agent which is usually an effective detergent they have superior cleaning power to other polishes. The waxes most commonly used are beeswax, carnauba wax and paraffin wax or ozokerite, and occasionally natural or synthetic resins, the harder waxes counteracting the stickiness of the beeswax. Silicone oils are often used to produce a water repellent finish. Care has to be taken to use a suitable polish for the surface to be treated. A great deal of modern furniture has a synthetic resin finish which does not require polishing.

Floor polishes have a similar formulation to furniture polishes although the wax content is designed to withstand much tougher wear and frequent wetting. Also a high gloss is desirable without slipperiness, this being achieved by using a substituted proportion of high melting point wax such as paraffin wax and ozokerite. Turpentine, white spirit, decalin, pine oil and dipentene are used as solvents, perfumes being commonly added to mask the odour of the solvent and produce a 'clean' polish smell.

Floor polishes must clean the floor surface and form a thin hard layer which does not attract further dirt and can be easily wiped

Polish manufacture—wax is blended in the upper tanks and is diluted with solvents in the tanks on the ground floor

clean. Liquid waxes can be thought of as paste waxes with a lower solids content which can be spread with a cloth or applicator. They are usually used in floor polishing machines and are particularly suitable for spraying. The difficulty here is to ensure that an even layer of product is dispensed over the floor and to avoid solvent separation in the polish reservoir. Water-based emulsion type floor polishes are often termed 'self-polishing' or 'dry-bright'. Although these have been in use for many years in the Scandinavian countries, USA and Canada, they have only recently been popularized in the UK. They were discovered accidentally in the 1920's during experiments with soap emulsified mixtures of water and carnauba wax.

It was found that the thin milky emulsions formed dried out when applied to a surface to give a thin, continuous, glossy film. The toughness of the film was enhanced by the addition of shellac dissolved in ammonia or morphaline. Later a high shellac/low wax formulation was used with an aqueous borax solution as the emulsifying agent. Finally, modern self-shine polishes were formulated which contain alkali soluble resins instead of shellac, together with emulsion polymers and synthetic emulsifiable waxes. On spreading self-shining polish over a floor the solvent evaporates and deposits the emulsified wax particles in a very thin layer of about 2 μm thickness. The dimensions of the wax particles are so small that the film reflects light almost as well as a continuous mirror surface.

Water-based polishes must not be used on surfaces such as bare wood or cork which could swell and discolour. They are quite safe to use on man-made floorings such as vinyl, thermoplastics, rubbers and bitumens which might be damaged by the solvents used in the traditional paste polishes. Occasionally there is a build up of wax on floors which have been polished over long periods. It is necessary to remove this to prevent imbedding of dirt and an uneven surface. If the floor is not damaged by solvents a white spirit cleaner is usually used. Otherwise alkaline detergents enable the wax film to be softened and then washed away.

Self-shining polishes normally have to be used on a clean surface as they have no detergent properties. Recently, however, liquid preparations have been introduced which clean and polish floor surfaces. These two-in-one polishes are water/wax emulsions. They break down on application to yield an aqueous detergent solution which can be mopped off with the surface soil while the wax layer adheres to the floor.

Motor car polishes are similar to furniture polishes except that a small quantity of a mild abrasive such as kieselguhr, china clay, Fuller's earth, chalk or titanium dioxide is added to remove 'traffic film'. The use of silicone oils as additives is widespread, producing a surface which repels water and forces it into small discrete globules which quickly evaporate.

Shoe polishes contain hard waxes such as montan wax, paraffin wax and carnauba wax, with a solvent which is usually turpentine or white spirit. Oil soluble dyes are added to the polish for colouring purposes, or in the case of black polish, finely divided carbon is used. Glycerol is usually added as a moistening agent (humectant) and to prevent fungal growth, together with metal soaps such as zinc or aluminium stearate which facilitate application. An interesting physical feature of paste shoe polishes is the appearance of rings on the cooled surface due to the crystallizing out of the myricyl alcohol present in the carnauba wax. Shoe creams are water/wax emulsions which also contain small quantities of solvent such as stearin, and emulsifying agents such as triethanolamine.

NON-WAXY POLISHES AND CLEANING AGENTS

Metal polishes clean and brighten metal surfaces by removing tarnish and greasy dirt. Essentially they are mixtures of mild abrasives such as Fuller's earth, jeweller's rouge (red iron oxide), chalk or aluminium silicate, with a fatty acid or fatty ester and ammonia, alcohol and water. Occasionally detergent and water-softening materials are also added to facilitate the removal of the tarnish.

'Instant dip' type silver cleaners are usually based on thiourea solutions in dilute hydrochloric or sulphuric acid. The unpleasant smell of the thiourea is often masked by small amounts (0·1–0·3%) of an odiferous material such as benzaldehyde. Solutions of sodium or potassium cyanide can also be used to clean silver by dipping, but their highly toxic nature makes them unsuitable for domestic use.

Glass cleaners contain an alcohol base such as iso-propanol with a small amount (5%) of ammonia and a synthetic detergent such as 'Teepol'. To this is added a non-scratch abrasive such as whiting or talc. Silicones, colour and perfumery additives are often used and occasionally insecticides such as DDT. Antimisting preparations are usually mixtures of a humectant such as glycerol with a metal soap and an abrasive such as titanium dioxide.

Heavy duty 'waterless' hand cleaning preparations for removing oil, tar and greasy dirt contain a blend of fats such as stearine and lanolin in deodorized paraffin (kerosine). This is stirred into water containing an emulsifier, such as triethanolamine, at about 70 °C and on cooling it sets into a clear gel. Pine oil or terpineol is often added to mask the smell of the ingredients and also to act as a disinfectant.

8

Scouring powders contain a detergent powder and an insoluble abrasive such as pumice, feldspar or talc, together with builders such as bleaches and perfumes. Lavatory cleaners consist mainly of sodium bisulphate which is prevented from caking by the addition of a little pine oil and also contains a little detergent of the alkyl aryl type such as DDB (dodecyl benzene).

Chapter 4

Cosmetics

For thousands of years it has been the custom of human beings of both sexes to decorate their bodies with coloured pigments and other materials. During the last century the skill of the chemist has been called upon to extend the scope of cosmetic preparations and to improve their formulation. Cosmetics for the hair and face have attracted particular attraction as the focal points for person to person contact, but a host of other preparations has appeared on the market during the last few decades. The rapid growth of the cosmetics industry has promoted intensive research into the nature of emulsions and surfactants and there have been important developments in the use of acetoglycerides and thixotropic fatty acid esters. It has also been necessary to make detailed physiological studies of those parts of the body likely to be affected by cosmetics such as the lips, nails, hair and skin. Three main groups of cosmetic preparation seem to have emerged which will be described in this chapter.

(a) Make-up preparations which are intended for decoration and the concealment of blemishes. These include lipsticks, eye shadow, nail lacquers, mascara, rouge, powder and hair dyes.
(b) Materials for cleansing and protecting the skin, hair and teeth. In this category are lotions, skin creams, shampoos, toothpastes, barrier creams and 'fresheners'.
(c) Treatment preparations for the skin and hair such as depilatories, deodorants, sun-screening agents, hair-waving fluids and after-shave lotions.

Changes in fashions and advances in manufacturing technique have also produced variations in the application of cosmetic products. Thus the older types of vanishing creams gave way to pancake type foundation creams, which in turn have been replaced by the modern semi-liquid emulsions which flow easily and are simple to use. The introduction of dispensers such as the aerosol, plastic puffer and polythene ball applicator has also revolutionized the application of many types of cosmetic preparation.

HISTORY OF COSMETICS

Examination of the tombs of the ancient Egyptians has revealed that the use of cosmetic paints by women was already well established as long ago as 5000 BC. Eye decoration was common practice in the days of Queen Cleopatra, the lids and lashes being blackened with a galena paste, and the underparts being painted green with a naturally occurring copper compound. In Roman times it was fashionable for women to wear rouge and lip salve, and to treat their hands and nails with henna. It was about this time that the Greek physician Galen made what was probably the first cosmetic emulsion by stirring rosewater into a mixture of olive oil and molten beeswax. The Bible contains several references to the use of cosmetics by Jewish women —sometimes with disastrous consequences (4 Kings 9: 30–33). Kohl was also widely used for darkening the eyelids in eastern countries.

The Babylonians used pumice-stone to smooth and whiten their skins. The use of face powder made from powdered marble, borax and starch first became popular in England during the reign of Queen Elizabeth I. During the period of Cromwell's rule the use of cosmetics was frowned upon, but with the accession of Charles II the practice reappeared and enjoyed an even greater popularity. By 1770 the use of cosmetics among all classes had become so'widespread that it was thought necessary to pass an Act of Parliament to protect men from being led into matrimony under false pretences by the use of '. . . scents, paints, cosmetic washes, artificial teeth, false hair, Spanish wool, iron stays, hoops, high-heeled shoes, and bolstered hips. . . .' The Act seems to have had little effect, however, on the use of these aids to beauty! Today the production of cosmetics is a major industry catering for a wide variety of tastes and needs throughout the world.

EMULSIONS AND THEIR PRODUCTION

Since a large proportion of cosmetic formulations are in the form of emulsions, it is essential for the cosmetic chemist to be acquainted with this type of system. Emulsions are produced by mixing together two immiscible ingredients (phases) in the presence of an emulsifying agent in such a way that one of the phases (the disperse phase) remains stably dispersed throughout the other (the continuous phase) in the form of tiny separate droplets. These droplets can be readily seen under the microscope and usually vary in diameter from about 0·2 to 50 μm. Thus an emulsion is intermediate in nature between a suspension and a colloid. In the case of cosmetic emulsions one of the phases is always aqueous and the other an oily or lipid phase.

If the fatty phase is continuous the emulsion is known as a water-in-oil system (W/O type), while an oil-in-water emulsion (O/W type) has a continuous aqueous phase. Occasionally complicated multi-phase emulsions are produced in which the disperse phase is itself an emulsion. Hydrophilic (water-loving) colloid-forming substances, such as sodium carraghenate and sodium carboxymethylcellulose, produce thick glue-like solutions in water and are often used to thicken the aqueous phase. They also serve to 'protect' emulsions from small amounts of electrolyte to which they are often sensitive. Emulsions of the type described are opaque white milks or creams, but by reducing the particle size of the disperse phase to colloidal proportions (0·05 μm diameter) clear micro-emulsions are produced.

To form a lasting emulsion it is not sufficient to agitate mechanically or stir the phases together. As the disperse phase is broken down into tiny droplets there is a large increase in free surface, which makes

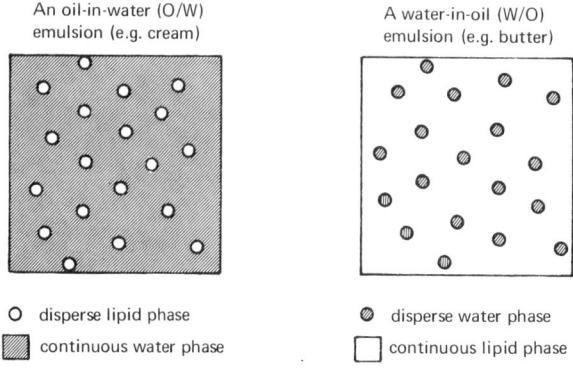

Fig. 4.1   Diagrammatic structure of emulsion types

the system unstable. Thus as soon as agitation ceases the droplets repeatedly coalesce in order to reduce the area of contact with the continuous phase. This goes on until two continuous layers of liquid reform.

In order to stabilize an emulsion and prevent it separating out ('creaming'), it is necessary to reduce the free surface energy. At the turn of the century Quincke, Donnan and others showed that this could be achieved by the use of surface active materials such as soaps. The stabilizing action of these emulsifiers is due to the formation of complex molecular clumps or micelles. Molecules of the emulsifying agent pack around the droplets of the disperse phase to form a continuous envelope, thus considerably lowering the surface tension between the two phases. In addition, this enveloping layer is usually electrically charged (zeta potential), and the mutual repulsion of the droplets is an important reinforcement of the stabilizing effect. The sensitivity of emulsions to even minute quantities of electrolyte is due to the leaking away of the zeta potential which results. The creaming of emulsions can also be hindered by increasing their viscosity, or by reducing the particle size of the disperse phase. Emulsions of high viscosity are not, however, usually permissible in cosmetic preparations.

The type of emulsion formed is governed mainly by the relative proportions of the two phases, the nature of the emulsifier and the method of preparation. With cosmetic emulsions the percentage weight of the disperse phase varies from 5 to 60% although up to 80% disperse phase is possible with some W/O emulsions. This is termed the phase volume fraction and is usually expressed as a decimal. Thus a PVF (ϕ) of 45% is written as 0·45. Inversion of an emulsion sometimes occurs when the phases 'change places' and an O/W emulsion becomes a W/O emulsion or vice versa. The formation of butter from cream is an example of inversion. This can be caused by a number of factors such as a change in phase concentration or temperature, or the presence of cations such as Al^{3+}, Cr^{3+} or Ni^{2+}. Emulsions are sometimes designed to invert on application to the skin or hair to render them more effective.

Over 3000 different substances can act as emulsifiers. Most of those used in the preparation of cosmetic emulsions, however, are either of the anionic or non-ionic type. The well known anionic soaps, substituted amine bases and aryl alkyl sulphates are mainly employed in the preparation of O/W systems. The more versatile non-ionic fatty acid esters and their polyoxyethylene derivatives can be used to prepare either type of emulsion. The type of emulsion which an emulsifier is likely to promote is indicated by its relative affinity for oil and water. This is known as its hydrophilic–lipophilic balance (HLB value). A widely used system allots a numerical HLB value to all emulsifying agents. Thus a surfactant molecule containing predominantly oil-loving (lipophilic) groups will be oil soluble and will have a low HLB number from about 3 to 6. Emulsions of this kind are required for the production of W/O type emulsions, or for the coupling of water soluble materials to an oily base. Wetting agents have a less pronounced lipophilic tendency and are found in the HLB range from 7 to 9. An HLB of 10 denotes a balance between lipophilic and hydrophilic groups and suitable emulsifiers for O/W systems are usually found aʰove this in the range 10 to 18, although several O/W emulsifiers have an HLB number as low as 7. Detergents are usually found in the HLB range from 13 to 15 although here again some excellent detergents are found in the 11 to 13 range.

Blends of emulsifiers are usually more effective than individual substances, and the HLB of an emulsifying blend is easily obtained by averaging the individual HLB values. Thus a blend of 75% emulsifier *A* (HLB = 5) with 25% emulsifier *B* (HLB = 15) would give a blend with an effective HLB of 7·5. Similarly, blends of ingredients can be allotted a composite HLB number in order to determine the emulsifier type required. For instance, a cosmetic cream may contain a lipid phase of 80% mineral oil, 15% cetyl alcohol and 5% lanolin. The HLB for this blend of ingredients would be obtained in the following way:

mineral oil	$80/100 \times$ HLB(10) =	8
cetyl alcohol	$15/100 \times$ HLB(15) =	2·25
lanolin	$5/100 \times$ HLB(12) =	0·60
		10·85

Thus the emulsifying system required for this blend will probably have an HLB value in the 10 to 12 range.

The behaviour of an emulsion largely depends upon whether it is

HLB	Substance
17	oleic acid
16	pine oil, ricinoleic acid
15	cetyl alcohol
14	castor oil, lauryl alcohol
13	nitrobenzene, ethyl benzoate
12	lanolin, carnauba wax
11	methyl silicone oil
10	mineral oil (paraffin oil), paraffin wax
9	beeswax
7–8	petroleum jelly

Approximate HLB factor for producing O/W emulsions of some common cosmetic ingredients

of the O/W or W/O type. The determination of emulsion type is therefore of great importance to the cosmetician. Often predictions can be made by consideration of such factors as the HLB value of the emulsifier, or the phase volume fraction.

It is often necessary, however, to carry out a more accurate diagnosis of emulsion type. Perhaps the simplest way of doing this is to mix small samples of the emulsion in turn with drops of oil and water respectively. The emulsion only mixes homogeneously with the liquid comprising the continuous phase. Another common test is to treat a sample of the emulsion with a finely powdered mixture of oil soluble and water soluble dyes of different colours. The colour which 'bleeds' reveals the nature of the continuous phase. Commercially a test lamp is used which relies upon the fact that only if the continuous phase is aqueous will an emulsion conduct electricity. Below the handle of the lamp is a split wire probe which is dipped into the emulsion. On switching on the electric current the lamp only lights up in the case of an O/W emulsion.

Preparation of Emulsions

The preparation of a cosmetic emulsion takes place in three stages. Initially the ingredients are heated to form a coarse emulsion. This is cooled and homogenized. The homogenization process produces a disperse phase of fine regular droplets, giving a high quality appearance, smooth feel and high degree of stability.

Before mixing, the ingredients are normally heated to a temperature of about 70 °C. This liquefies any solid components and reduces the viscosity to facilitate stirring. Phase mixing can be carried out simultaneously in continuous emulsification processes. It is more usual, however, to add the continuous phase to the disperse phase. This naturally causes an inversion of the emulsion at some point during the mixing process, and has been shown by experiment to give a high quality product. Stirring is usually continued during the cooling process; this ensures homogeneity especially if there is a high wax content.

Emulsification is completed by subjecting the cool liquid to some kind of shearing action to reduce the size of the disperse phase. For low viscosity emulsions this is usually carried out by high speed stirring with propeller or turbine type agitators. Portable mixers are often used which can be clamped to the top of the container. These are fitted with twin propellers of opposite pitch which are fixed to a vertical shaft driven at 800–1500 rpm by an electric motor. The lower propeller forces the liquid upwards against the downward thrust of the upper propeller, thus ensuring thorough stirring. If the emulsion

Mixers used in the production of cosmetic emulsions

is of high viscosity slower stirring must be used to prevent the entrainment of air bubbles. In this case anchor type stirrers are used which fit closely to the sides of the container.

To produce very fine emulsions of superior quality, colloid mills are employed. In these the liquid is violently sheared between two circular plates revolving at a speed of 1000–3000 rpm. The plates are either smooth or grooved and are set from 0·025 to 0·25 mm (0·001–0·01 in) apart. An alternative design which is very efficient involves pumping the liquid at high pressure through a narrow orifice against a spring loaded valve.

Recently ultrasonic vibrators have been used to produce fine emulsions and to incorporate solid additives such as pigments.

COSMETICS AND THE SKIN

In order to be able to formulate effective and safe products, it is essential for the cosmetician to know something of the structure and functions of the skin. This important barrier between the body and its environment performs the dual function of keeping out unwanted materials and micro-organisms, and retaining valuable substances such as water. It has a large surface area of some 1·8 m^2 (1·5 yd^2) and accounts for about one sixth of the body weight of an adult.

The skin consists of a thick bed layer known as the dermis which has a well developed system of blood capillaries. Upon this is a second layered region known as the epidermis which has no blood supply and is nourished by materials diffusing from the underlying dermis. The active region of the epidermis is the basal layer which is adjacent to the dermis. Cells are continually being produced in this layer, especially during sleep, and as they age they keratinize and clump together in layers to form a horny outer surface to the skin which is rubbed off and replaced.

Continuous with the epidermis are the hair follicles and the sebaceous and sweat glands. The epidermal surface is normally covered by the secretion products of these glands. This helps to maintain the moisture balance of the underlying tissues. The eccrine sweat glands produce a clear watery exudation which is usually acidic, and contains about 1% dissolved solids, mostly urea and sodium chloride.

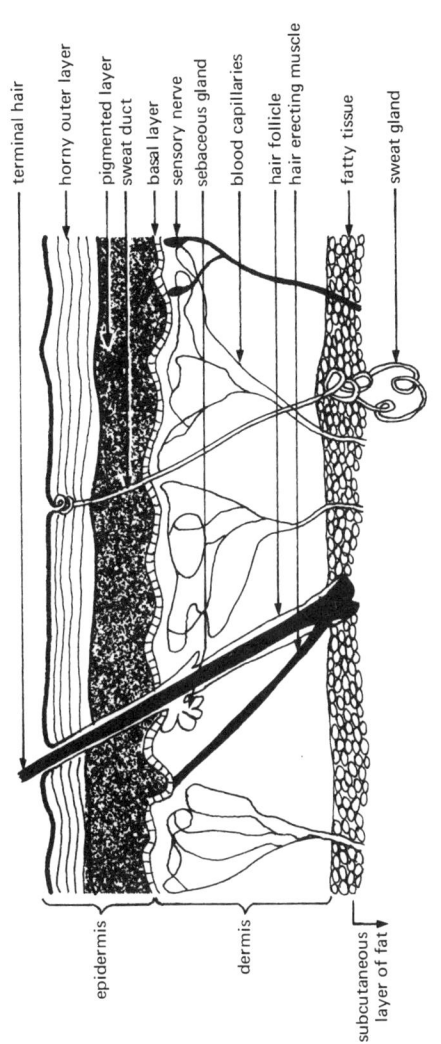

terminal hair
horny outer layer
pigmented layer
sweat duct
basal layer
sensory nerve
sebaceous gland
blood capillaries
hair follicle
hair erecting muscle
fatty tissue
sweat gland

epidermis

dermis

subcutaneous
layer of fat

Fig. 4.2   Diagrammatic section of human skin

Sweat produced from the apocrine glands is a whitish secretion containing a complex mixture of sugars, proteins and lipids. The decomposition of apocrine sweat by micro-organisms produces the unpleasant smell of body odour. The sebaceous glands produce an oily substance known as sebum, which is a mixture of fatty acids, esters, cholesterol and a variety of paraffins and other hydrocarbons. The sebum covers the epidermis and hair with a thin film which guards the skin against infection and reduces water loss.

The colour of the skin is due partly to the blood. The dermis and the epidermis also contain pigments, the most important of which is a dark brown compound called melanin. The amount of melanin in the skin is responsible for racial colour differences and is secreted in special cells (melanocytes) found in the basal layer of the epidermis. It is formed by enzymatic oxidation of the amino acid tyrosine, and its production is stimulated by two hormones which have been labelled α-MSH and β-MSH (MSH — melanin stimulating hormone). Melanogenesis also occurs in the presence of sunlight and other sources of ultraviolet radiation. Sun-screening preparations such as p-amino benzoic acid, transmit the tanning radiation which lies on the waveband between 300–390 nm (3000–3900 Å), and filter out the shorter waves which cause sunburn. Artificial skin-tanning agents usually contain dihydroxy-acetone which combines with the epidermal keratin to form a brown pigmentation. This effect is only temporary and does not protect the skin from sunburn.

Normally there is a continuous loss of water from the skin surface as insensible perspiration. If the water loss becomes excessive, the surface of the epidermis becomes dry and rough and loses its flexibility. This is called 'chapping' and in extreme cases causes fissures in the skin and bleeding. Chapping results from the removal of the surface fats through excess washing or prolonged exposure, especially in cold, windy weather. It is a condition which becomes increasingly common with age.

Although water loss can be reduced by the use of fatty materials rubbed on the skin, experiments have shown that dry, hard skin can only be softened by absorption of water. A few fatty materials such as cholesterol and lanolin are able to penetrate the skin, and possess both lipophilic and hydrophilic properties which make them useful skin conditioners or emollients. Ageing of the skin is thought to be due to a progressive lessening of its moisture-holding capacity together with keratinization changes due to the uptake of Ca^{2+} and Mg^{2+} ions. It is clear from this that to be effective, a skin preparation should contain both an oily and aqueous phase. Emulsions are ideal in this respect, the water they contain softening the skin and the lipid fraction forming a moisture seal. It is claimed that the use of humectants such as glycerol and polyethylene glycol permits absorption of atmospheric moisture, but there is no firm evidence on this point.

Recently, wide use has been made of the iso-propyl esters of stearic, palmitic and linoleic acid as emollients. Not only are these materials absorbed by the skin, but they are not greasy or sticky, which makes them easy to apply. In addition they are useful in modifying the properties of the traditional vegetable and mineral oils to suit different skin conditions.

CREAMS AND LOTIONS

There are many different types of cosmetic cream in common use, each having individual characteristics suiting it for a particular purpose. They may be broadly divided into preparations in which the emulsion itself produces the required effect, and those in which the emulsion acts as a convenient vehicle for an active ingredient of some kind. It is possible to 'tailor' an emulsion to produce a specified product by varying the nature and properties of the ingredients and making a careful choice of emulsifying agent. Usually O/W systems are used for formulations which have a low fatty content such as hand creams and cleansing lotions. For products with a higher fatty content such as cold creams and emollients either type of system can be used. The choice in this case is usually determined by other factors such as production and packaging costs.

The oil phase nearly always contains mineral oil to which is added a fatty ester to reduce greasiness and lower viscosity. Waxes and fats such as stearic acid, cetyl alcohol and lanolin are used to produce an emollient film on the skin and to thicken the oil. Thickening is also carried out by the use of metal soaps such as aluminium stearate. The aqueous phase can be thickened by the addition of hydrophilic colloids such as gums or alginates, and contains water soluble ingredients such as bactericides and emulsifiers. For the production of a liquid emulsion HLB values are of great assistance in choosing a suitable emulsifier. For solid creams, however, HLB values for the emulsifier can only be obtained experimentally, not by calculation.

Face creams and lotions are used as emollients. They prevent the skin from becoming rough and coarse due to excessive moisture loss under unfavourable conditions, and restore its natural flexibility and smoothness. Occasionally rejuvenating agents such as oestrogenic hormones are included, but opinion is divided as to their effectiveness. Both O/W and W/O systems can be used with a high fatty content to produce a rich, soothing product which is easy to apply. This type of product is often applied by women as a massage cream before retiring at night, or as a soothing cream for sore or chapped skin. It can be used generously, a slight greasy residue being permissible. The oil phase contains substances such as lanolin, mineral oil, petroleum jelly, olive oil and sometimes smaller quantities of more unusual and often exotic sounding materials such as avocado pear oil and raisin seed oil. Unsaturated oils which are likely to become rancid are protected by an antioxidant such as butylated hydroxytoluene. Both ionic and non-ionic emulsifiers, such as glyceryl monostearate, borax, cholesterol and wool-wax fatty alcohols are commonly used. Quaternary ammonium compounds such as cetyl trimethyl ammonium bromide are less popular because of their likelihood of causing skin irritation.

Cleansing creams are designed to remove soil from the face, including make-up and skin debris. To be effective they have to be thixotropic to facilitate dispensing and application. A thin film must be produced on the face which can readily be removed by swabbing with a paper tissue. Both oil and water soluble soil must be removed with one application, the skin being left freshened and clean, and covered with a fine emollient film. Most cleansing creams are O/W type emulsions which are more effective at removing traces of make-up than soap and water. The oil phase is the most effective part of the emulsion, and some creams have been marketed which contain no water at all. Typically a thin type of mineral oil is used which is thickened and stabilized with a thixotropic wax such as stearic acid, and contains a small quantity of an emollient ingredient like spermaceti or cetyl alcohol.

As face powder will not adhere satisfactorily to dry, clean skin it is necessary to use a *foundation cream* to provide a suitably receptive surface. O/W type emulsions are commonly used for this purpose, to produce a slightly sticky, non-occlusive matt film which ensures the adherence of an even layer of powder. Pigments or opacifying agents are sometimes included to 'set off' the colour of the powder.

Vanishing creams can be used either as foundation creams or hand creams. They owe their name to the almost invisible, non-greasy matt film which they leave on the skin after application. Stearate soaps which are produced during the mixing process are used as emulsifiers—sodium stearate forming a firm cream and the potassium derivative a softer one. The desired sheen and consistency is produced after cooling by carefully controlled milling. Drying and crust formation in the container is prevented by the inclusion of a humectant such as glycerol.

Hand creams and lotions are commonly used by housewives to keep the skin of their hands in good condition and prevent chapping, in spite of domestic tasks which require prolonged immersion in hot detergent solution, or produce heavy soiling. Most preparations of this type are lightly perfumed O/W emulsions which can be quickly rubbed into the skin without leaving a sticky or wet feeling, and form a thin barrier film on the skin surface. In order to achieve this the lipid phase is usually composed of 5 to 20% of a waxy material such as cetyl alcohol or stearic acid plasticized for ease of application with a thixotropic fatty acid ester. Small amounts of emollients and silicone oils are also often included. The aqueous phase contains a humectant such as sorbitol or propylene glycol and occasionally a mild antiseptic. Triethanolamine soaps are commonly used emulsifying agents, although specially prepared surfactant blends such as 'Span' and 'Tween' (mainly laurate, stearate and oleate esters) are rapidly gaining popularity.

Industrial barrier creams are used to protect the hands of work people from special types of soiling encountered during their employment.

An extremely popular type of *emollient 'all purpose' cream* which is widely used in European countries is based on a W/O emulsion with a PVF of about 0·6 ('Nivea' cream).

In preparations where the emulsion is used as a vehicle for an active ingredient, due regard must be paid to the compatibility and mode of action of the additive. Thus a hormone cream containing an oestrogen such as pregnenolone acetate must be able to penetrate the skin surface to be of use. A sun-screening cream on the other hand must remain on the surface.

Sun-screening agents can be incorporated into lotions, oils or creams, which should have an emollient effect to prevent sun drying, and be resistant to removal by sea bathing or rubbing off on clothes. The active ingredient is usually included in the oil phase as an opaque powder or an organic substance which selectively absorbs ultraviolet radiation. A sun-screen index is often used, based upon the transmission spectrum of a 1% concentration of the material using a light wavelength of 308 nm (3080 Å).

Deodorant and antiperspirant creams are other types of product which use an emulsion vehicle, although they are also produced in

An ointment filling machine capable of filling 60 tubes a minute

Screening agent	Sun-screen index
ethyl *p*-dimethyl aminobenzoate	14·8
methyl umbelliferone	7·7
p-amino benzoic acid	7·4
heliotropine	6·5
sodium *p*-amino salicylate	4·3
methyl salicylate	4·0
3-carboxy coumarin	3·0
resorcylic acid	2·2

the form of sticks and lotions with an alcoholic base. Deodorants contain a bactericide such as hexachlorophene which prevents decomposition of the secretion of the apocrine sweat glands which is responsible for socially unacceptable body odour. About 20% of an antiperspirant is normally added to prevent perspiration in localized areas such as the armpits. Aluminium salts such as aluminium chlorohydrate have proved very effective for this purpose. For ease of application a polythene roll-ball or spray applicator is often used.

Powders and talcs. The most popular and widely used of all cosmetics is *face powder*, which is used to protect and improve the appearance of the facial skin. Face powders are formulated to produce an even matt layer which will cover skin blemishes and absorb surface moisture and oils, thus preventing skin shine. The weathering effect of sun and wind is also reduced. Colouring pigments and perfume are usually added to make the product more acceptable. The preparation has to adhere well to the skin without producing a 'chalky' appearance, and should not be readily removed by friction or rain. In addition it must not be so absorbent as to produce skin drying, and must be smooth and easy to apply.

A judicious blend of several ingredients is necessary to produce a satisfactory product. Thus talc and zinc oxide are used for smoothness, titanium oxide and precipitated chalk for covering power, kaolin and magnesium carbonate for absorption, and metallic soaps such as magnesium stearate for adherence and waterproofing. Particle size is also an important factor in producing a smooth, high quality powder. High speed pulverizers reduce the particle diameter to below 20 μm. A highly efficient machine for this purpose is the

Powder blending machines

Large scale powder blending machines

micronizer, which spins the powder at very high speed in a shallow cylindrical container using tangentially directed jets of air. By this means a particle size of less than 10 μm can be produced without any appreciable temperature rise or risk of metallic inclusions. Perfume is added before pulverizing but is first absorbed in an equal weight of magnesium carbonate to facilitate mixing. Suction techniques are used to fill the powder containers because of the difficulties involved in handling such finely divided material.

The inclusion of a small quantity of binding agent such as a fat or mucilage enables a compressed powder cake to be prepared which can be packed into a flat metal case. The degree of compression is carefully controlled to enable the powder to be easily removed with a powder puff without crumbling the cake. Care also has to be taken during manufacture to avoid the inclusion of bubbles of air which cause the product to 'flake'. Compacts of this type first became popular in the late 1930's.

Moisture absorbent talcum powders are prepared from the mineral talc to which a mild astringent such as zinc oxide is added. About 0·5% of a deodorizing bactericide such as hexachlorophene is usually included. If perfume is to be added this is first absorbed in magnesium carbonate as described above. The product can be applied from a box as a dusting powder, but is more usually dispensed from sprinkler top cans. Recently talcum powder dispensers in the form of aerosol sprays have been marketed.

COLOUR AND ITS USE IN COSMETICS

With the exception of the creams, almost all cosmetic preparations include colouring of some sort. Great care has to be taken in the selection of the colorant, not only must this produce the required effect under the conditions of usage, but it must be innocuous. A list of recommended colouring agents has been published by the Society of Cosmetic Chemists (*J. Soc. Cos. Chem.*, **13**, 379 (1962)). There is no legislative control of the use of colouring materials in cosmetics in the UK, although in 1957 a bill was passed restricting the use of colour in foodstuffs. In the USA, however (and in Canada), there is a list of 116 'certified colours' laid down in the Food, Drug and

Cosmetics Act 1938. These colouring materials are divided into three groups, i.e. those for general use in food, drugs or cosmetics (F.D. & C.), drugs and cosmetics only (D. & C.) and those only to be used externally, except around the eyes and mouth (External D. & C.). There is strict control on purity and the use of heavy metal derivatives. The list of certified colours was amended in 1960 and again in 1963 and 1966 in order that new colours could be provisionally included.

A number of organic dyestuffs which are soluble in either water, oil or alcohol find application in a wide range of cosmetic materials such as lipsticks and rouges. Two of the most important types are the soluble diazo and acid dyes, such as Orange 2 (D. & C. Orange No. 4), and xanthene dyes such as the substituted fluoresceins (D. & C. Red No. 21). The solubility of these dyes is often a serious disadvantage, however, and there is a much greater demand by the cosmetician for insoluble colouring materials.

Insoluble lakes can be prepared from the soluble dyes by precipitation on a suitable substrate such as aluminium hydroxide. Carmine is a natural colorant which is used in the form of a lake in this way. Coloured pigments such as insoluble diazo dyes are widely used in the form of suspensions. In addition a number of insoluble inorganic pigments such as iron oxide ochres, umbers and siennas are used. An interesting group of synthetic pigments are the ultramarines which can be obtained in a variety of colours ranging from greeny-blue to pink, by heating together a mixture of sulphur, silica, soda, Glauber's salt, resin and china clay. This material is equivalent to the natural, rare and fabulous material, *lapis lazuli*.

Metallic powders can also produce novel effects in certain cosmetic preparations such as eye shadow. Powdered aluminium and bronze are most commonly used either in their natural colours or dyed with a basic dyestuff using tannic acid as a mordant. Another special effect is obtained by the use of pearl essences which are prepared either from fish scales or synthetically from bismuth oxychloride.

Four types of cosmetic preparation which rely on colour for their effect are lipstick, rouge, eye make-up and nail lacquer.

Lipsticks consist of a moulded fatty base containing soluble stainers and a suspension of coloured pigments. The base is made from a thixotropic blend of waxes and oils which enables an opaque layer of colour to be easily and smoothly applied. Good adherence of the film to the lips is important to avoid unsightly staining of teacups and cigarettes. Also the stick must be strong enough to allow a clear firm line to be drawn without danger of snapping. An unusual cosmetic requirement of lipstick is that the ingredients must be edible in small quantities and have an acceptable taste, in addition to being non-irritant to the lips themselves. Careful formulation is necessary to ensure that 'sweating', 'blooming' and colour changes do not occur after prolonged storage over a wide range of temperatures.

The traditional ingredients of a lipstick base are beeswax and castor oil, but if only these materials were used the stick would 'drag' and have a dull appearance. Small amounts of hard waxes such as candelilla and carnauba improve the gloss and impart strength to the stick, while microcrystalline waxes such as ozokerite are useful for oil retention. Lanolin is also a common ingredient and contributes a valuable emollient effect to the product. The tendency of lanolin to drag and feel greasy can be counteracted by the use of acetoglycerides and iso-propyl fatty acid esters such as iso-propyl myristate. Diethyl sebacate and propylene glycol are used as blending agents and solvents for the stainers. Oxidation can cause serious deterioration in a lipstick producing unpleasant rancid odours and taste, in addition to possible discoloration and the formation of toxic products. This can be largely prevented by the use of antioxidants such as butylated hydroxytoluene (BHT), nordihydroguaiaretic acid (NDGA) and butylated hydroxyanisole (BHA). BHA is often blended with other antioxidants, as this has been found to make them more effective (synergistic effect).

The lipstick is manufactured by thoroughly milling together the coloured and oily ingredients. Originally this was carried out by repeatedly passing the mixture through rollers. In modern factories a colloid mill is often used. This consists of two metal or carborundum discs with an adjustable gap which can be reduced to as little as 25 μm. One disc remains stationary while the other revolves at speeds of up to 5000 rpm as the coloured material is passed between them. When the colour is fully dispersed the resulting paste is stirred in a heated pan containing the molten waxes and other components.

'BHA'
butylated hydroxyanisole

'BHT'
butylated hydroxytoluene

'NDGA'
nordihydroguaiaretic acid

Mixing is performed carefully using a slow speed, propeller type mixer in order to avoid the entrainment of air bubbles. Perfume and flavour are added to the mixture immediately prior to casting, which is carried out using an automatic or split mould. Often the moulded sticks need to be 'flamed' by passing them quickly through a gas flame or along an infrared heating tunnel. This causes slight remelting of the surface and produces on setting a glossy blemish-free appearance.

Rouge is intended to be used sparingly beneath other preparations such as creams or powder. The colouring materials and other ingredients are similar to those used in the formulation of lipsticks. Liquid rouge contains humectants such as sorbitol or propylene glycol together with a thickening agent such as one of the cellulose ethers to prevent sedimentation of the pigment. Preservatives such as methyl *p*-hydroxybenzoate also have to be included. Cream rouges are based on a mixture of petroleum jelly, mineral oil and a thixotropic ester such as iso-propyl myristate, hardened with a microcrystalline wax. Some cream rouges are also available which are either O/W or W/O emulsions.

Eye make-up is probably one of the oldest types of cosmetic, and it has recently become very fashionable for both daytime and evening wear. It is not desirable to use many of the organic type of colouring agents in the vicinity of the eyes, and usually highly purified natural pigments are used. To produce special iridescent effects small (1%) amounts of aluminium dust or fish scale extract are used. Eye shadow is usually prepared in the form of a cream, liquid emulsion or stick. The formulation of the sticks and creams must be aimed at producing a highly thixotropic product which will not drag the delicate skin around the eye. This is achieved by the use of fatty acid esters and microcrystalline waxes. Mascara is often manufactured in the form of a flat cake, although creams dispensed from squeeze tubes are also available. In both cases the colour paste is stirred into a water/wax mixture in the presence of an emulsifier such as triethanolamine stearate. The preparation should be easy to apply, drying quickly to give a hard smear resistant film which will not crack, or cake the eye lashes together.

Nail varnish is used to produce a tough glossy film on the finger-nails, which can be coloured or clear according to taste. This film must not peel or chip easily when typing or carrying out other office tasks or household chores. Neither must it be affected by contact with water or detergent solutions. The lacquer must be easy to apply and quick drying without showing brush marks. It must be viscous enough not to drip off the brush or run to the edge of the nail. The colouring (if present), and other ingredients, must not affect the nails or the skin either on the hands or other parts of the body such as the face which are touched by the hands.

The most commonly used film-forming ingredient is nitrocellulose. The addition of aryl sulphonamide formaldehyde resin or polyester resins improves both the gloss and the adherence of the film to the nail. Plasticizers such as dimethyl phthalate are also used to prevent cracking and chipping. Commonly used nitrocellulose solvents are ethyl acetate and methyl ethyl ketone (MEK) which are often coupled with alcohols such as butanol to increase their solvent effect. Nail varnish colours must be bright, opaque and resistant to the action of detergents and other common household fluids. Lakes and pigments are widely used in addition to iridescent ingredients such as 'pearl essence' (guanine crystals from fish scales). Colour matching is usually carried out by eye, comparison being made between lacquer films coated on sheets of opaque white glass. By running the lacquer down a vertical glass sheet and timing the period of tackiness, the drying time can be determined. Adhesion power is tested by coating a sheet of glass with the lacquer and then determining the resistance of the hardened film to removal by a razor blade. Other tests designed to find the resistance of the lacquer film to water, abrasives and scratching are also carried out.

Nail varnish removers used to contain a large proportion of acetone but this is often omitted in modern formulations. The acetates of butanol, ethanol and 1-pentanol (amyl alcohol) are usually blended with a small quantity of an emollient oil to produce an effective stripping liquid.

PREPARATIONS FOR THE HAIR

Hair is built up from long chain protein molecules known as keratins. It grows at an uneven rate from narrow slanting pits in the skin known as follicles. Proliferation of the follicular hair cells is very rapid, the length of the hair fibre increasing at the rate of about 1 cm (0·4 in) a month. The cortical cells of the fibre are surrounded by a layer of transparent overlapping scales which form the cuticle. The colour of the hair is due to the presence in the cortex of egg-shaped granules of the pigment melanin. These lie end-to-end in rows parallel to the longitudinal axis of the fibre. It has been found that heavily pigmented hair fibres are more resistant to alkalis, protein-splitting enzymes and reducing agents, than white hairs.

The keratin molecules which form the hair fibres are built up by the condensation of large numbers of α-amino acids. The condensation always involves a carboxyl group from one molecule and

$$\underset{\text{glycine}}{\overset{\overset{\displaystyle NH_2}{|}}{CH_2COOH}} + \underset{\text{alanine}}{\overset{\overset{\displaystyle CH_3}{|}}{NH_2CHCOOH}} \xrightarrow{\text{condensation}} \underset{\text{dipeptide}}{\overset{\overset{\displaystyle NH_2 \qquad CH_3}{| \qquad \quad |}}{CH_2CONHCHCOOH}}$$

an amino group from the other. Peptide linkages are formed which are a characteristic of protein molecules. From the examples shown below it would appear that the polypeptide molecule has a zig-zag structure, but work by Pauling and Corey in 1950 showed that the molecular form was almost certainly a special type of spiral known as an α-helix.

Amino acids occasionally bond at right angles to the axis of the main polypeptide chain forming short side chains. Diamino compounds such as lysine form side chains with a basic amino group, while dibasic amino acids form side chains which possess a carboxyl group. Amino acids such as leucine produce inert hydrocarbon side chains.

basic side chain	NH₂(CH₂)₄CH—COOH H₂N	(lysine)
acidic side chain	HOOCCH₂CH—COOH NH₂	(aspartic acid)
inactive side chain	H₃C CHCH₂CH—COOH H₃C NH₂	(leucine)

axis of
polypeptide chain

When the acidic side chain of one polypeptide lies so that it is adjacent to the basic side chain of another, transfer of the acidic hydrogen occurs to form what is known as a salt link. These salt links

—COOH NH₂— ⇌ (aqueous acid/alkali) —COO⁻ ⁺NH₂—
                                                  H

| acidic side chain | basic side chain | adjacent polypeptide chains joined by a salt link |

(represents main axis of polypeptide chain)

are destroyed by dilute acids and alkalis. This may be demonstrated by steeping hair in a 0·1 N solution of hydrochloric acid for 20 minutes, when it will be found to stretch more easily. On thoroughly washing with distilled water to remove the acid, the links are reformed and the hair regains its original properties.

The cystine molecule can be thought of as two α-amino acid units joined 'back-to-back' by a short bridge of two sulphur atoms. This unusual structure allows the molecule to be incorporated in two different polypeptide chains, the latter being linked together by the disulphide bridge. Most of the 4% sulphur present in hair keratin is accounted for by these linkages, which are very sensitive to alkaline hydrolysis and reducing agents. This fact is of great significance in the formulation of hair-waving preparations.

$$\begin{array}{c} NH_2 \\ | \\ SCH_2CHCOOH \\ | \\ SCH_2CHCOOH \\ | \\ NH_2 \end{array}$$

cystine

cystine linkage between
adjacent polypeptides

In addition to salt links and disulphide bonds there are two other forces which tend to hold adjacent polypeptide chains together. When the carbonyl oxygen atom of one chain lies very close to the imino hydrogen atom of another, a weak link is formed between them called a hydrogen bond. Molecules which are very close together are also attracted by what are known as Van der Waals forces.

Both these bonding forces are very small, but exist in such numbers as to exert a considerable overall effect. When hair is wetted, water molecules are able to infiltrate between the polypeptide chains, thus considerably weakening the hydrogen bonds and Van der Waals

(*a*) hydrogen bonding between
adjacent polypeptides

(*b*) weakening of hydrogen bonds by
inclusion of water molecules

attraction. This effect is increased under alkaline or acidic conditions, or if the temperature is raised. The bonds reform on drying unless excessive separation or disalignment of the fibres has occurred. *Waving and setting* the hair involves the breaking and subsequent resetting of the different types of inter-chain bond. The most active and widely used cold-waving agents are alkaline solutions containing thiols, such as ammoniacal solutions of thioglycollic acid (mercapto-ethanoic acid). A disadvantage of the thiols is their powerful and unpleasant odour which is difficult to mask. The few thiols which are odourless, such as lauryl mercaptan, have no waving effect.

$$HSCH_2COOH \qquad C_{12}H_{25}SH$$
thioglycollic acid lauryl mercaptan

Pretreatment of the hair with a reducing agent causes rupture of some of the disulphide bonds (scission) and thus enhances the effect of waving solutions. Oxidizing agents such as hydrogen peroxide have the reverse effect. This explains the difficulty in 'setting' peroxide-bleached hair.

The breakdown of disulphide links with thiols involves a series of complicated reactions which have been the subject of much study. The initial stages of the breakdown series yield mercaptans and possibly sulphenic acids, although the formation of the latter has not been confirmed.

sulphur bridged polypeptides thiol   mercaptans   disulphide

Hydrolytic breakdown may continue to give unsaturated vinylidene and aldehydic side chains accompanied by the liberation of hydrogen sulphide.

Since the breakdown of the disulphide linkage is the result of attack by thiol anions (RS⁻), then clearly the effectiveness of a

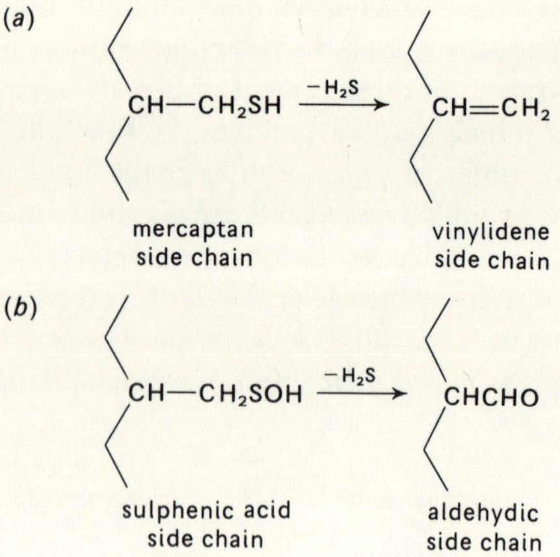

(a)

$$CH-CH_2SH \xrightarrow{-H_2S} CH=CH_2$$

mercaptan   vinylidene
side chain   side chain

(b)

$$CH-CH_2SOH \xrightarrow{-H_2S} CHCHO$$

sulphenic acid   aldehydic
side chain   side chain

particular thiol as a hair-softening agent will depend upon its degree of ionization. The effectiveness of thioglycollic acid is due to its molecular structure which favours ionization; cysteine on the other hand, which does not ionize so readily, is a poor hair-softening agent.

$$\overset{\text{SH}}{\underset{\text{CH}_2\text{COOH}}{|}}$$

thioglycollic acid

$$\overset{\text{SH NH}_2}{\underset{\text{CH}_2\text{CHCOOH}}{| \ \ |}}$$

cysteine

Thioglycollic acid is prepared commercially as the sodium salt by reacting sodium chloroacetate in aqueous solution with sodium hydrogen sulphide.

$$NaSH + \overset{\text{Cl}}{\underset{\text{CH}_2\text{COONa}}{|}} \longrightarrow \overset{\text{SH}}{\underset{\text{CH}_2\text{COONa}}{|}} + NaCl$$

sodium hydrogen   sodium   sodium
sulphide   chloroacetate   thioglycollate

Prior to waving, the hair is first shampooed and then wetted with the softening solution before winding on to small plastic curlers. The modern croquignole type of curler has a narrow waist to prevent the lock of hair bulking in the centre. The hair is held in position by a small clip or band. The tip of the lock of hair is often wrapped in a small slip of tissue to facilitate handling. More recent styles call for a much looser wave which is obtained by winding wider locks of hair on large plastic or aluminium 'rollers' having a diameter of about 25 mm (1 in). This is a much quicker method, but the wave produced lasts for a shorter time—usually about five weeks.

The strength of the alkaline waving solution is such that the hydrogen bonds and salt links are inactivated and about 20% of the disulphide bonds rupture in a period of from 15 to 30 minutes. At the end of this time the softened hair is thoroughly rinsed, and then hardened using a 'neutralizer' which is an oxidizing agent such as sodium perborate or sodium bromate. Reformation of sulphide and disulphide bonds takes place in a complex series of reactions of the following kind:

(a)

$$C{=}CH_2 + CH_2{-}CH\text{(SH)} \xrightarrow[\text{sodium bromate}]{\text{oxidation}} CH{-}CH_2SCH_2{-}CH$$

reformed sulphide link

(b)

$$CH{-}CH_2\text{(SH)} + CH_2{-}CH\text{(SH)} \xrightarrow[\text{sodium bromate}]{\text{oxidation}} CH{-}CH_2SSCH_2{-}CH$$

reformed disulphide link

The neutralizing process requires patience, and if the curlers are removed before oxidation is complete, the hair may still be partially softened and the wave fail to 'take'.

Experimental trials of new waving lotions or techniques are usually carried out using a 'half-head' method. The hair of the subject is parted down the centre, and the test wave carried out on one side only. The other half of the head is treated with a standard process which can be used as a control for comparison. It is important to remember that the individual polypeptide chains are so minute in comparison to the hair fibres, that only a very small degree of chain curvature is required to effect quite pronounced waving of the hair.

Before the use of thiols, alkaline solutions containing sulphites and borax were used to soften the hair by bond scission. Hot-waving, in which solutions of this kind are used, requires the application of heat to the wetted curls by either chemical or electrical means. Cold-waving has largely replaced hot-waving because of its relative simplicity of operation.

Hair can be temporarily set by simply softening the hair with water. This weakens the hydrogen bonds as shown above. Wetting the hair does not affect the disulphide bonds, however, and reversion to the original form is rapid, especially in a moist atmosphere. In a dry atmosphere, such as the American 'Middle West', hair which has been water-waved remains 'set' for a considerable time.

The tightly crimped hair of coloured women can be straightened by exactly the same processes as those used to wave hair. The waving lotion is used to soften the hair which is then combed or brushed out to release the tight curls.

Chemical analysis of hair-waving fluids is usually carried out by titration of the alkaline component against a standard acid solution using methyl orange as an indicator. The thioglycollic acid content is then determined using a standardized iodine solution. The reaction in this case is represented by the following equation:

$$\underset{\text{thioglycollic acid}}{\overset{\overset{\displaystyle SH \quad SH}{\displaystyle | \quad \; |}}{HOOCCH_2 + CH_2COOH}} \xrightarrow{+I_2} \underset{\text{disulphide derivative}}{\overset{\overset{\displaystyle S\text{---}S}{\displaystyle | \quad \; |}}{HOOCCH_2 \; CH_2COOH}} + 2HI$$

Shampoos are used to remove greasy dirt, dandruff, sweat and unwanted odours such as cooking smells and tobacco smoke from the hair and scalp. It is desirable that the detergent used should be efficient in cleansing the hair in this way without leaving it dull and unmanageable. Unfortunately, apart from their powerful wetting action which adversely affects the set of the hair, the detergents in common household use roughen the hair cuticle. This causes loss of lustre and produces electrostatic 'fly' on brushing or combing. These effects are not produced on washing the hair in grease solvents such as trichloroethylene, so that earlier theories which blamed poor hair condition after washing on the removal of the sebum from the hair seem to be incorrect.

Other detergents have an irritant effect on the skin or eyes, or produce too little (or too much) foam. The most satisfactory types have been found to be polyethylene glycol derivatives of the fatty oils obtained from the palm kernel and the coconut, such as sodium lauryl polyethylene glycol sulphate. Most shampoo formulations are prepared as liquids or creams. The liquids are invariably based upon one of the lauryl sulphate derivatives, to which is often added a persistent bactericide, such as hexachlorophene, to produce so-called medicated shampoos. The addition of a variety of other substances such as egg, lanolin and beer seems to have little effect other than as a sales stimulus. Certain alkanolamides, however, such as coconut monoethanolamide have a useful conditioning effect on the hair.

Liquid cream shampoos for 'dry' hair are produced by thickening the detergent base with an insoluble metallic soap such as magnesium stearate. This is usually formed in situ by the interaction of sodium stearate with an appropriate metal salt. The use of transparent plastic sachets to package the shampoo enables the attractive pearly lustre due to the suspension of metallic soap to be seen by the purchaser. Further thickening by materials such as the methylcelluloses produces a shampoo cream which can be dispensed from a metal squeeze tube.

Powder shampoos based on sodium lauryl sulphate are now seldom used because of the greater convenience of liquid shampoos. The 'dry' shampoos used by hairdressers are usually mixtures of high boiling petroleum ethers (Stoddard solvent), together with one of the chlorinated hydrocarbon solvents such as tetrachloroethylene.

In addition to its role in the washing and waving of scalp hair, water is also used in shaving. Although our ancestors were mostly bearded, crude razors have been found dating back to well before the Roman Empire. As the design of razors improved and the use of soap became widespread, shaving became more popular, and the 'cut-throat' popularized by Louis XI in the fifteenth century was widely used in England by the middle of the seventeenth century. It is interesting to note that the reason given by Alexander the Great for ordering his troops to shave was so that they could not be grasped by the enemy during battle!

Shaving mugs appeared about 1800 and the first patent for a shaving stick was taken out in 1903. Lather shaving cream was first sold in jars in the UK just after World War I and the first aerosol shaving creams appeared in 1952. The use of the electric razor has steadily increased in the last decade, but a large number of men still prefer 'wet' shaving especially since the introduction of the stainless long-life razor blade.

Shaving soap enables the beard to be rapidly wetted and softened, and in addition acts as a skin lubricant to facilitate the action of the razor. To be effective the soap must produce a good firm lather which is innocuous and non-corrosive to the razor, and can be readily rinsed away. The required product is a balanced blend of about 80% fatty acid soaps with a small amount of free fatty material, together with about 8–10% glycerol and the same quantity of water.

Shaving creams are prepared by including a high proportion of potassium or 'soft' soaps in their formulation. The fatty acids used are carefully chosen and their saponification controlled to give a lustrous crystalline structure to the product which is usually dispensed from a squeeze tube. Superfatting materials such as lanolin are usually added to give an emollient effect.

Brushless creams and aerosol shaving creams on the other hand have no wetting effect on the beard and their application must be preceded by washing with soap and water. They are usually O/W emulsions in which sodium stearate is used to disperse the lipoid phase, which consists of mineral oil with added superfatting ingredients. The function of a brushless cream is, as its name suggests, to remove the need for a shaving brush. In addition it provides a good lubricant film for the razor and enhances the softening effect of the water. Glycerol is usually added as a humectant, and thickening agents such as sodium alginate and polyvinyl pyrrolidine are used to stabilize and give body to the emulsion. Menthol is also widely used because of its pleasant freshening effect.

Hairdressings are used by both sexes to fix their hair during the day, and to improve its lustre. The early types of men's hairdressings were in the form of pomades, prepared from soft animal fat such as bears' grease perfumed by impregnation with essential oils. With the development of the petroleum industry, mineral oils and petroleum jelly enabled the formulation of the modern hair creams and brilliantines. Liquid brilliantines are often light mineral oils which are suitably coloured and perfumed, but are unpopular because of their 'greasy' nature. One way to reduce the oil content is to use a 20% solution of a suitable oil, such as iso-propyl myristate or castor oil in ethanol. But this is inevitably accompanied by a loss in fixative power. Petroleum jelly hardened with a suitable wax, and coloured and perfumed, can be used as solid brilliantine. Transparent solid brilliantines are composed of mineral oil which has been gelled with a metallic soap such as magnesium stearate.

The emulsion type of hair creams remain by far the most popular of men's hairdressings, especially for dark hair. These can be either O/W or W/O emulsions, although the latter emulsions are more popular. Products of this type have to be carefully formulated to ensure that the emulsion breaks down immediately on rubbing into the hair. If this does not occur the cream remains visible as an unsightly white mass which gives the hair a dull greasy look. O/W emulsions readily break down on application as the aqueous phase is rapidly removed by evaporation and absorption into the hair. The problems involved in the formulation of W/O emulsions are more difficult to resolve. The emulsion in this case must be stable enough to resist the mechanical shocks and thermal changes encountered during transport and its shelf life, but must break down immediately on application to the hair. This is achieved by preparing an insoluble metal soap in situ to form a fragile but solid 'skin' around the droplets of the disperse phase. The emulsion is then prepared by high speed stirring alone, since homogenization would break the protective skin and cause creaming. The perfumes used in these delicate emulsions have to be carefully selected since they are often surface active and might seriously affect the emulsion. The oil phase in hair creams is a mixture of mineral oil and fats and waxes, such as stearic acid, cetyl alcohol and beeswax. Common emulsifiers are borax, triethanolamine and sorbitan sesquioleate. The proportion of the aqueous phase varies from between 50–65%, although in W/O creams this sometimes rises to 70%.

Hairdressings used by women are mainly required to gloss the hair, and must contain a minimum of water as this adversely affects waving. Usually a light brilliantine or O/W emulsion with a high oil

10

content is used, which can be applied by spraying. Hair conditioners are also used as rinses after shampooing. These contain cationic detergents such as Cetrimide B.P. (a mixture of cetyl and myristyl trimethyl ammonium bromide) which are preferentially taken up by the acidic side chains of the polypeptides of the hair fibres. These are retained even after rinsing and contribute a sleek gloss to the hair and inhibit electrostatic 'fly'.

Dyestuffs are often used on the hair either to mask white or grey hair, or to produce a decorative effect. The older types of dyes, such as camomile and saffron, merely coated the surface of the hair fibre with a coloured layer. The more permanent penetrating dyes are of more recent origin. A major problem involved in the dyeing of the hair is that the process must be carried out at body temperature. This means that for effective penetration to occur, the dye must have a small molecular volume and a minimum of side chains.

picramic acid
(orange-red)

4-nitro
1,2-phenylenediamine
(yellow)

Two common hair dyes

Dyes of this type are rarely soluble and so have to be used as suspensions. Hence the term disperse dyes is used to describe them. A popular method of application is by means of a colour shampoo, which combines a detergent with an appropriate dye. Another way of overcoming the problem of penetration is to use small molecules which readily penetrate the hair fibres, and which can then be induced to link up within the fibre using an oxidizing agent such as hydrogen peroxide. This produces large molecules of dyestuff permanently embedded in the hair structure. Unfortunately many people seem to be acutely allergic to compounds such as *p*-phenylenediamine which are commonly used in this type of oxidation dyestuff.

Allergy and Product Testing

Allergy is an unusual reaction exhibited by an individual in circumstances in which normal non-allergic persons are unaffected. Effects commonly produced are swelling, nettle rash, reddening and blistering of the skin, asthma and watering of the eyes. This is thought to be due to the release of histamine or histamine-like substances from the tissue cells. A wide range of sensitizing allergens exists and is being continually extended by the use of new plastics, drugs, dyestuffs, insecticides, fibres and the like. Certain chemical substances, which are termed haptens, are able to combine with serum or tissue protein to produce allergens. It is these materials which cause contact dermatitis and other unpleasant effects when applied to the skin. An interesting example of this effect is the sensitization of individuals to tetramethylthiuram disulphide after wearing rubber shoes in which this material has been used during manufacture. If such persons use a soap in which the disulphide is used as an antiseptic, extensive dermatitis and skin blistering results. Obviously such materials must be excluded from cosmetics as far as possible.

To test the reaction of the skin to possible allergens, or to detect the presence of such substances in cosmetic preparations, a technique known as patch testing is used. A convenient area of skin, usually on the upper arm or thigh, is first cleaned by swabbing with a little ether. A small gauze pad about 25 mm (1 in) square is then impregnated with the test material and laid on the skin. A larger piece of plastic film is used to cover the patch and fixed firmly in position with adhesive tape or plaster. A control patch of clean gauze is fixed alongside. If itching or discomfort occurs within an hour, the patches are removed immediately and the skin rubbed with an antihistamine cream. Otherwise the corners of both patches are lifted after 24 hours and the skin examined. If there is no sign of reddening or other change, the patches are replaced and left for a further 24 hours. The skin is then carefully re-examined for signs of reaction.

A positive reaction is obtained when inflammation, and in severe cases blistering or even ulceration, occurs at the test site but not at the control. The inflammation is almost always accompanied by severe itching, and may last for several days. Occasionally delayed effects may appear a week or more after the test. A new cosmetic product is first tested on a few volunteers, and then if no effects are evident, upon 200 or so other individuals. The product is usually rejected if a positive reaction is obtained during the first 400 tests.

Allergy to lipstick is not usual, but is sometimes caused by the eosin stains used. It is not clear why this should be so, but it is thought to be due to the presence of impurities or tiny particles of undissolved solid. Women who exhibit allergy on using standard lipstick are able to use special varieties in which the eosin content is very low. Lipstick allergy is also thought to be caused in certain cases by the perfume used. Citrus oils and synthetics which are ketones or aldehydes are usually avoided for this reason.

Other coloured cosmetic products which are liable to cause allergy are nail varnish and hair dyes. The widespread allergy of persons to such dye bases as *p*-phenylenediamine has already been mentioned, and its use is banned in France, Austria and Germany.

In addition to allergy tests, the use of many novel synthetic materials in modern formulations has led to the need for toxicity tests on new cosmetic products. This is a difficult task because the action of such materials as carcinogens may not be apparent for many months or even years. Tests must also be carried out to prevent the damage or irritation of areas adjacent to the place of application which may become accidentally affected. An example is the Draize 'rabbit-eye' test which is used to estimate the effect of shampoos which accidentally enter the eye.

Cosmetic products are also subjected to stability tests to detect possible damage by micro-organisms, emulsion breakdown, chemical decomposition, colour fading, perfume deterioration, precipitation or settling. Exhaustive laboratory and consumer tests are then carried out to evaluate the performance of the product.

Chapter 5

Perfumery

'And the Lord said unto Moses, Take unto thee sweet spices, stacte [styrax], and onycha [labdanum], and galbanum; these sweet spices with pure frankincense . . . and thou shalt make it a perfume.'

Exodus 30 : 34

HISTORY OF PERFUMERY

The art of the perfumer is founded deep in history. The ancient Egyptians used sweet smelling resins to embalm their kings, and objects from the tomb of the mighty Tutankhamen still retain traces of perfume after 3000 years. Odiferous smokes or incenses, produced by burning aromatic gums and barks, were also used by the Egyptians for religious purposes. At Heliopolis the sun worshippers burnt incense at sunrise, myrrh at noon and kaphi at dusk. It is from these practices that perfume owes its name as it is derived from the Latin *per fumum*—through smoke.

There are many references in the *Iliad* and *Odyssey* of Homer to the use of perfumed materials by the ancient Greeks and Romans. The first recipe book for compounding perfumes was written by an ancient Greek, Theophrastus, who was born in 370 BC. 'Perfumers seek upper rooms', he writes, 'which do not face the sun, but are shaded as much as possible. For the sun or a hot place deprives the perfumes of their odour.'

At this time it was common for men to wear perfume, and to scent their hair, beard and even their wines. Rose and violet seemed to be most widely used and were favourites of the Emperor Nero, who is reputed to have burnt a whole year's supply of incense at the death of his wife Poppaea. Caesar on the other hand abominated perfume and said he preferred men who stank of garlic to those who smelt of perfume. Napoleon also disliked perfume although his wife, the Empress Josephine, used such quantities of musk that her private boudoir still smells of it to this day. Musk was also used to perfume the mortar during the building of Arabian mosques—and is still discernible in some cases.

Myrrh and frankincense were highly treasured in the past for ceremonial and religious purposes and it is no accident that they were chosen, together with gold, for the gifts of the wise men to the infant Jesus. Frankincense means 'easily ignited' and is a gum found oozing from the bark of a tree growing in the Himalayas and northern Arabia. It is still the main component of incense used in churches today, just as it was in the time of Moses.

The Koran also makes mention of odiferous materials and plants such as musk and hyacinth (*Sura* 55).

The perfumes and extracts of certain flowering plants were thought to possess medicinal and prophylactic properties. The Chinese believed musk to be a cure for epilepsy and other ills, and even today the Chinese peasant will often superstitiously slip a fragment of musk under the nail of his big toe to ward off snakebite. In Britain too it was long the custom for a bunch of fragrant flowers to be carried ahead of a high court judge as a nosegay to ward off the dreaded gaol fever. The nursery rhyme 'ring-a-ring of roses' refers to the belief that the Plague could be avoided by screening oneself with a 'pocketful of posies'—not very effectively it seems, as the sneeze was a symptom of Plague and 'all fall down' tells its own story.

As early as the ninth century AD it is believed that Arab chemists had succeeded in distilling plant oils, and a number of perfumes were brought back to Britain by knights returning from the Crusades in the twelfth century. The discovery of alcohol in the fourteenth century was of great significance to the perfumer. It proved an ideal solvent for the many odiferous materials which by then had begun to flood in from all parts of the world. During the next two hundred years a number of works on the subject of perfumery appeared, many of which were concerned with the production of aqueous flower extracts such as lavender-water, and perfumed ointments and vapours. A number of such recipes appear in Baptista Porta's *Natural Magick* published in 1658. In Book Two we read—'Anoynt the Pill of Citron or Lemmon with a little Civet; stick it with cloves and Races of Cinnamon: Boyl it in Rose-water, and it will fill your chamber with an Odoriferous fume.'

It is not surprising that these valuable commodities should be the subject of dishonest dealings and Porta ends his chapter with a warning '... these perfumes are often counterfeited by Imposters:

wherefore I will declare how you may discern and beware of these Cheats: for you must not trust whole Musk-Cods, there being cunning Imposters, who fill them with other things—Goats blood a little rosted, or toasted bread; so that three or four parts of them beaten with one of musk will hardly be discovered'.

A typical 'flower-water' was 'Neroli Oil', an orange oil extract which was introduced by the Duchess of Neroli. It was used in 1709 by Johann Maria Farina as one of the chief ingredients of his Kölnisches Wasser, or as it is more usually known, eau de Cologne—which still retains its place today as a successful perfume.

The Renaissance was a stimulus to interest in perfumes and cosmetics and by the end of the eighteenth century the cultivation of flowers for the extraction of essential oils was already a thriving industry in the South of France. The small town of Grasse established itself as the world centre of the floral perfume industry, and floral perfumes were soon to replace the colognes in popularity. These in turn were later to be influenced by the discovery of the synthetic odiferous materials resulting from the great upsurge of organic chemistry in the latter half of the nineteenth century.

Thus the heavy floral 'notes' of perfumes were modified with synthetic esters, ketones and aldehydes such as methyl nonyl acetaldehyde, amyl salicylate and later methyl phenyl carbinyl acetate. Just before World War II perfumes containing 'woody', 'green' or 'animal' notes such as sandalwood, musk and civet were popular, although, as behoves a nation of gardeners, the British have always favoured floral perfumes such as lavender, rose and violet. After the war, in the 1940's, there was a swing back towards the 'powdery' notes such as vanillin and orris root. For men's toilet preparations heavy spicy notes were used, reminiscent of the perfumes of Spain and South America.

Today the perfume industry is of great importance, producing odiferous materials not only for personal use, but for a wide range of commercial products including polishes, detergents, inks, disinfectants, air fresheners and novelty cards.

SMELL PERCEPTION

In order to be detected an odiferous substance must be volatile. It must also be water soluble to pass through the aqueous film covering the smell receptors. In addition many odiferous materials are soluble in the fatty lipid coating of the sensory nerve endings. Two types of olfactory receptor are embedded in a pair of yellow patches of tissue lying in the nasal cavity just above the roof of the mouth. The majority of the receptors are nerve endings of the olfactory nerve. These terminate in a small tuft of about six sensory hairs which project into the air passages. The other type of fibre is a thread-like ending of the trigeminal nerve. The exact mechanism involved in the detection and analysis of an odour is not known, but impulses are received by the olfactory bulb of the brain via the nerve endings and these are interpreted by the higher centres as smell. As the nasal cavity is open to both nose and mouth, it is inevitable that the sensations of smell and taste are intimately connected.

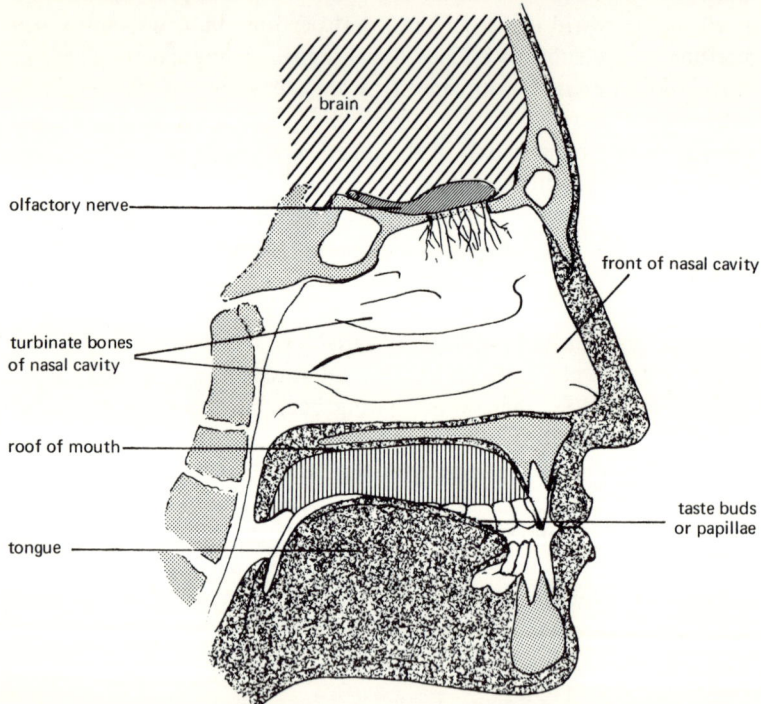

Fig. 5.1   Section of the human skull showing the structure of the olfactory mechanism

A great deal of research has been undertaken to determine the way in which different odours are appreciated by the olfactory system. In the past some 30 different theories have been put forward, but none has stood up to experimental testing. One of the difficulties has always been the absence of any unit to measure the intensity of an odour, in the way in which light or sound can be measured. In addition there is the problem of identifying and classifying the very wide range of smells. At present this can only be carried out using the human nose. This inevitably involves individual human factors difficult to eliminate.

Experiments recently carried out in the USA by Amoore and others provide evidence to support a stereochemical theory of odour. This links the molecular shape of a compound with the nature of its odour. The germ of this idea had been proposed 2000 years before by the poet Lucretius and was revived in 1949. The first task was to divide all smells into a few primary odours, corresponding to the primary colours (red, blue and yellow) and the primary tastes (salt, sour, sweet and bitter). One classification suggested four possible elementary odour types, namely, fragrant, acid, burnt and caprylic. Another proposed no less than 18 distinct types. Amoore systematically examined some 600 odiferous organic compounds and concluded the existence of seven primary smells as shown.

Primary odour	Chemical example	Familiar example
Camphoraceous	camphor	rosemary
Musky	butyl dinitrotoluene	angelica
Floral	geraniol	roses
Pepperminty	menthone	spearmint
Ethereal	carbon tetrachloride	dry cleaning fluid
Pungent	acetic acid	vinegar
Putrid	butyl mercaptan	bad eggs

Amoore went on to propose the existence of specially shaped receptors in the olfactory membrane. Each of these would correspond to the shape of molecules producing the seven primary odours. Thus although benzaldehyde, nitrobenzene and cyclo-octanone have no obvious chemical links, they all possess a similar smell of almonds because their molecular shapes have certain similarities. This lock-

and-key theory seems to be supported by experimental evidence. For instance, synthetic molecules of a specific shape have been produced, and found to possess a predicted odour type. In addition, naturally occurring complex odours such as cedarwood, have been imitated successfully using mixtures of the seven primary odours.

Recently the existence of specialized receptor sites has been demonstrated experimentally. Using single isolated olfactory nerves from the frog, and delicate micro-electrodes, it was shown that the nerves were only stimulated by a particular odour.

One of the fascinating things about the sense of smell is the minute amount of material necessary to produce an odour response. The exact threshold of perception is difficult to estimate and varies with individuals; some people having virtually no sense of smell are termed anosmic. Others with a highly developed sense of smell can detect substances such as musk and citrus oil in amounts which are only of the order of about 0·1 microgrammes. Even this fact pales into insignificance when we consider the sense of smell of insects. Thus the male silkworm butterfly can detect the smell of its mate at distances of up to 19 kilometres (12 miles)! G. K. Chesterton comments on man's poor nose for smells in the 'Song of Quoodle':

> 'They haven't got no noses,
> The fallen sons of Eve;
> Even the smell of roses
> Is not what they supposes'

Another interesting fact is that single 'pure' odours are rarely found in nature. Many of the essential oils which produce the fragrant smell of flowers are extraordinarily complex mixtures. Thus lime oil contains at least 240 observable constituents. The analysis of such mixtures has been greatly helped in recent years by the application of new and powerful techniques such as gas chromatography, infrared spectroscopy and nuclear magnetic resonance.

It has been reported that this complexity of the natural essential oils is accepted as a desirable feature by the nose. This has led to attempts to 'randomize' synthetic mixtures by the use of radioactive bombardment in order to improve their odour. Although successful in certain instances on a small scale, there appear to be major problems involved in the application of this process on a commercial scale.

Clary sage grown for its essential oil content

PERFUMERY MATERIALS FROM PLANT SOURCES

Many plants produce oily substances which have little smell and are non-volatile. These materials such as olive and almond oil are known as fixed oils. In addition there is a large range of plant oils which are highly aromatic and volatile. These are known as essential oils and are found mainly in the flower of the plant although smaller quantities are also present in the fruit, stem, root and leaf. In addition there are a number of gums and resins produced in the bark of trees which produce odiferous substances called balsams. Such plant materials are invaluable in the production of perfumes, and much work has gone into the perfection of extraction methods.

The essential oils contain complex mixtures of organic compounds including alcohols, esters, aldehydes, ketones and terpenes. It is probable that these are waste products of plant metabolism and are used by the plant in the dual role of attracting fertilizing insects and perhaps repelling those which are destructive. The odiferous materials are probably formed in the chloroplasts of the leaf. Here they combine with glucose to form glucosides, which are transported around the plant structures. In certain regions, especially in the flower, enzymes are produced which attack the glucosides and regenerate the essential oils. An interesting experiment designed to show the loss of essential oil vapour from odiferous plant material such as flower petals has been described by Devaux. A piece of plant tissue is laid on a mercury surface which has been dusted with talc. Absorption of oil displaces the talc giving visual indication of perfume loss.

The most important ingredients of the essential oils are probably the terpenes and their derivatives. These have a basic structure which may be regarded as having been formed by the head-to-tail fusion of a number of isoprene molecules. Characteristic terpene-based perfumery ingredients are limonene, camphor, citronellol, geraniol and terpineol.

isoprene   limonene

The extraction of the essential oils from the parent plant material is carried out by three basic techniques. These are distillation, solvent extraction and expression—the first two being of major importance.

Distillation

This is the oldest and probably still the most widely used of all extraction methods. It is only suitable, however, for the essential oils which are not readily decomposed at steam temperatures. These include the oils of lavender, rose, peppermint and rosemary. In the older crude 'field' distillations the plant material was boiled with water and the oil separated from the mixed distillate. Serious losses of product resulted from this method due to ester hydrolysis and oxidation. Overheating invariably produced further undesirable side-products. In modern practice the plant material, with the exception of leaves and petals, is first chopped or crushed to facilitate extraction and then charged into a steam still. About a tonne of material is

Steam distillation
of essential oils

Distillation of essential oils (Long Melford)

treated at a time by injecting steam into the vessel or suspending the plant material on a horizontal sieve plate above boiling water. The vapours of oil and steam are then condensed using water cooled coils. Occasionally the process is carried out at reduced pressure. The mixed distillate flows into a 'flask' where the oil separates out as a layer above the water and can be removed. The small quantity of oil dissolved in the water is not wasted but either extracted with a solvent or sold as a water perfume such as lavender-water.

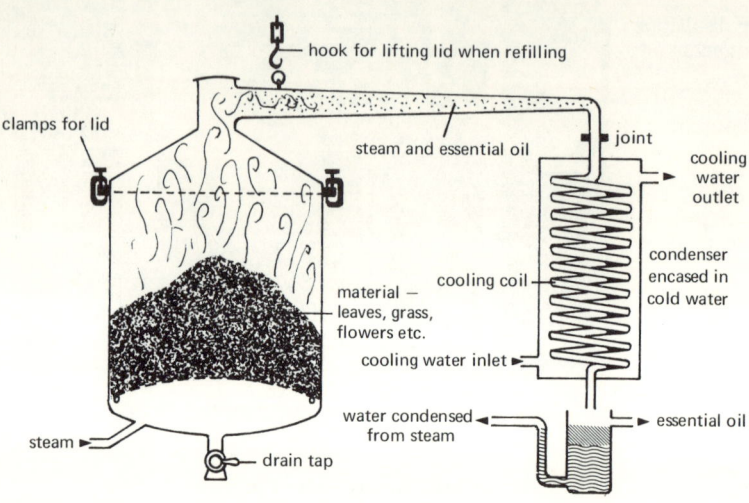

Fig. 5.2 A steam distillation unit for the extraction of essential oils from plant material

The exhausted charge which remains in the still is used either as fuel or cattle food. The high cost of maintaining this modern type of plant for intermittent use during the year can only be justified in a few special areas such as Grasse (France), Calabria (Italy) and the rose-growing areas of Bulgaria. The yield from distillation varies enormously according to the plant material being extracted. Lavender has a high extraction rate of about 1% while rose has a low yield of about 0·06%.

Solvent Extraction

The less drastic process of solvent extraction is more suitable for the less stable floral oils which would be decomposed by distillation. It was first used commercially by Millon in 1856 but was a failure owing to the loss of expensive solvent which occurred during the process. This was overcome by the use of a sealed extraction vessel when the process was successfully re-introduced about 1890. A number of solvents have been used such as chloroform, ether and alcohol, but the most successful to date has been petroleum ether. This is now used on a large scale.

The solvent is continuously percolated through a series of trays containing the plant material using a countercurrent technique, until extraction is complete. The liquid extract is filtered and passed to a vacuum still, where the more volatile solvent is distilled off at low pressure. The solid waxy residue containing the essential oils is called a 'concrete'. Since this contains plant waxes, pigments and other non-volatile materials it is of greater bulk than the equivalent product obtained by steam distillation.

The concrete is churned with ethanol for about 24 hours in a machine called a batteuse. This dissolves out the essential oils and allows the insoluble wax to be filtered off. A small fraction of the wax is soluble in alcohol and this is removed by lowering the temperature to about $-20\,°C$ and refiltering. Distillation of the alcoholic extract is carried out under reduced pressure to produce the 'absolute'. The absolute produced by this process often contains traces of colouring matter and these are removed by exposure to ultraviolet light or by co-distillation with ethylene glycol.

As with other processes the yield varies from one plant material to another, although the yield from solvent extraction is usually higher than that from steam distillation. Thus the yield of rose concrete is about 0·24% producing 0·12% of absolute, which is double the yield obtained by distillation with steam.

A recent development in the field of solvent extraction has been the use of liquefied butane gas under pressure, at a temperature of $-15\,°C$. This method is very suitable for the most delicate floral oils such as lilac and gardenia. The absolutes obtained by the process are of high quality and are called 'butaflors'. Liquid carbon dioxide has also been used in low temperature extractions of this kind.

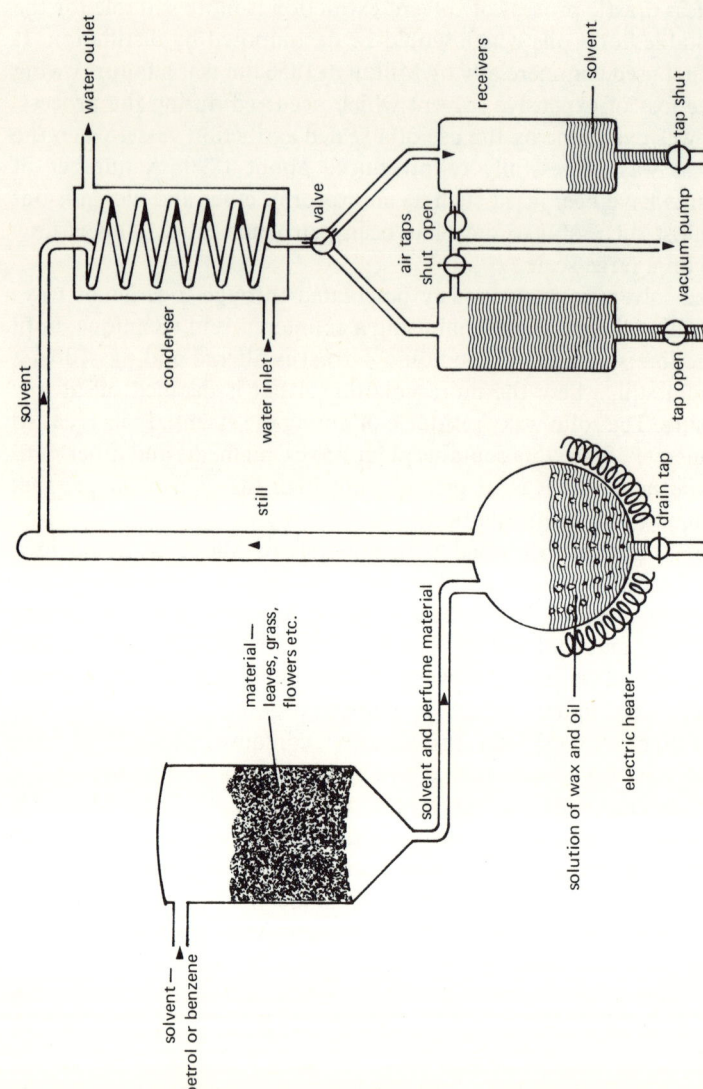

Fig. 5.3 Simplified diagram of an essential oil solvent extraction plant

An old fashioned type of solvent extraction process still in restricted use is called 'enfleurage'. This is particularly suited for the extraction of the oils of jasmine and tuberose because these continue to be produced for some time even after the flower has been picked. The

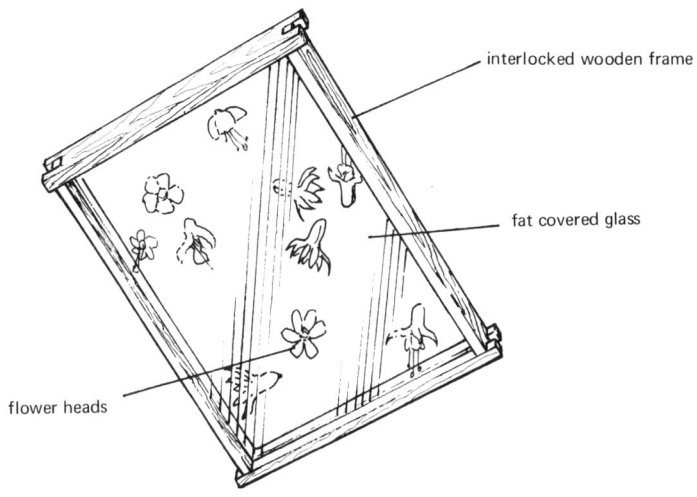

interlocked wooden frame

fat covered glass

flower heads

Fig. 5.4 Diagram of the fat coated glass sheet and frame used traditionally for recovery of essential oils by enfleurage

solvent is a specially purified mixture of fats which is spread in a thin layer on glass sheets called 'chassis'. The exact composition of the fatty mixture, which usually contains lard or beef suet, is often kept as a family secret. The flowers are lightly attached to the fat, and the plates stacked in piles for a period of from one to three days. The withered flowers are then removed and replaced by fresh blooms. This process is repeated for periods of up to two months until the fat is considered to be saturated with the floral oil. The enfleurage operations were originally all carried out by hand, but are now mechanized to a large degree. The perfumed fat produced by enfleurage is called 'pomade' and was used many years ago as a dressing for the hair. It is now treated with ethanol in a similar way to concrete in order to extract an 'enfleurage absolute'.

An alternative method using a fat solvent is termed maceration. This is used for most flowers except tuberose and jasmine which are treated by enfleurage for the reasons given above. The plant material is agitated in metal drums containing the molten fat at a temperature of about 65 °C. The oil-bearing cells are ruptured by this process and the odiferous material absorbed by the molten fat. After some hours, the spent flowers are removed from the fat by pouring through a screen, or in a more modern plant by centrifuging. The fat is then returned to the extraction, recharged with flowers and the process repeated. After the required degree of saturation, the fat is treated with ethanol and the absolute extracted as before.

Expression

The citrus oils are extracted from lemons, oranges, limes and bergamot fruit. The oil is secreted in small cells in the peel of the fruit. These cells are easily ruptured to release the oil, as can be seen by bending a piece of orange or lemon peel backwards. Hand pressing of the rind of citrus fruits is now dying out although it is still used in the production of orange and lemon oils. Expression of bergamot oil and large quantities of lemon oil is now carried out by machine.

Final Processing

Before the absolutes are used in compounding perfumes, further purification is carried out. Non-volatile material is removed by re-distillation at reduced pressure, and in certain cases fractional distillation is used to remove the more volatile components. These are highly prized by the perfumer, and command very high prices.

PERFUMERY MATERIALS FROM ANIMAL SOURCES

In addition to the materials of plant origin, there are a few animal products which have been used in perfumery for many centuries. The only four of commercial importance are ambergris, castoreum, musk and civet. Musk and ambergris are expensive and only used in the finest perfumes. The cheaper products are used on a large scale in the compounding of a wide range of less costly preparations. Alcoholic

Final distillation of essential oils

extracts of the substances are usually made and these are used in the form of tinctures.

The main function of animal products is as fixatives to stabilize the bouquet of blended perfume. This ensures that the more volatile components do not evaporate off before the others, the perfume thus remaining consistent over a long period. Plant materials such as oak moss and benzoin resin are also used as fixatives but they are inferior to the animal products. A number of odourless organic liquids of high boiling point, such as benzyl benzoate (b.p. 323 °C), have been used for the same purpose, but these have a 'flattening' effect on the bouquet.

An additional property of the animal fixatives is that they have a powerful odour. This is unpleasant in high concentrations but on considerable dilution imparts 'body' to a perfume which would otherwise be insipid.

The use of ambergris and civet was already well established by the tenth century AD. About this time it is recorded that the Sultan of Fez presented a neighbouring chieftain with a gift of one pound (0·5 kg) of each of these materials. Castoreum and musk do not appear to have been used before the seventeenth century, however.

The origin of the animal fixatives is interesting and unusual. Civet is produced as a foul smelling secretion of glands found in both sexes of the civet cat. The active component in this defensive liquid is civetone. In Africa the cats are caged and fed on a meat-rich diet. On being annoyed or frightened a stream of the secretion is ejected and collected. Musk is a glandular secretion of the male musk deer which inhabits the mountainous regions of India, Tibet and China. The active constituent in this case is a lactone called muscone. The Canadian beaver secretes castoreum in small glands or 'pods' which are possessed by both sexes. It has a characteristic powerful phenolic odour. The relative cheapness of the product is in part due to the high prices fetched by the beaver fur. Ambergris on the other hand is very expensive. It is a greasy substance which varies in colour from white to near black, and is found floating in the sea. It is produced by a pathological condition of the intestine of the sperm whale.

SYNTHETIC PERFUMERY MATERIALS

As with dyestuffs, the high cost of the natural odiferous materials stimulated research into ways of producing cheaper synthetic organic substitutes. The first big developments in this field coincided with the upsurgence of organic chemistry in the latter half of the nineteenth century. Perkin, the discoverer of the first aniline dyes, also prepared the first of the synthetic perfumery products, coumarin, in 1875. In the same year Tiemann and Reimer produced synthetic vanillin. A quarter of a century later Tiemann was to make another breakthrough in the synthesis of ionone, which has a powerful smell resembling violets, although it is not present in the flower itself. More recently Ružička was awarded the Nobel Chemistry Prize for his syntheses of muscone and civetone.

During the last few years synthetic routes have been discovered for the preparation of many other important perfumery chemicals such as citronellal and geraniol, previously only obtainable by the fractionation of essential oils.

Although the synthetics tend to be 'rough' compared with their natural counterparts, and also require skilful blending, they have the advantages of cheapness, consistency and large scale production. Several thousand synthetics are used in formulation research and many hundreds are in regular commercial use. The bulk of these materials are aromatic and terpene alcohols and their corresponding esters, aldehydes and ketones.

ALCOHOLS AND ESTERS

The aromatic and terpene alcohols are important sources of synthetic materials for perfumery. They blend well with other ingredients and have soft notes which can be modified and intensified by esterification. Thus while benzyl alcohol has only a faint aromatic odour, the acetate ester has a pronounced jasmine-like smell which has led to its widespread use in the preparation of floral perfumes and fruit flavours. It was prepared for this purpose in Germany during World War II by heating benzyl chloride with anhydrous sodium acetate.

benzyl chloride + NaOCOCH₃ (sodium acetate) → benzyl acetate + NaCl

Another ester of benzyl alcohol, benzyl benzoate, although possessing only a faint smell is a valuable fixative for jasmine and other floral perfumes. Cinnamyl alcohol is also a good fixative and has a pleasant hyacinth smell. It is produced synthetically by reduction of cinnamaldehyde, although its esters occur naturally in plants such as styrax.

cinnamaldehyde → cinnamyl alcohol (aluminium iso-propoxide)

β-phenylethanol has a smooth rose-like odour, and is used in floral perfumes together with the related phenyl propanol, and their acetate and propionate esters. Phenylethanol is prepared commercially by condensing ethylene oxide with benzene at low temperature.

benzene + ethylene oxide → phenylethanol (AlCl₃, 10 °C)

The terpene alcohols are derivatives of the terpenes which are reactive cyclic structures occurring widely in essential oils. One of the most important produced synthetically on a large scale from turpentine is α-terpineol. It has a pleasant, fresh, woody smell which is in demand as a lilac soap scent, and perfuming agent for disinfectants ('Dettol'). The turpentine is emulsified with sulphuric acid and stirred for some hours at 25 °C. Terpin hydrate is formed, which is separated from the unchanged turpentine emulsion, alkali-washed and then boiled with 0·1% sulphuric acid. Crude α-terpineol steam distils over from the mixture and is fractionally distilled at reduced pressure. Terpenyl acetate is used as a lavender substitute and the butyrate and formate are also used in perfumery.

Another important terpene alcohol is geraniol, which together with its geometric isomer nerol, has a pleasant rose petal smell. This is produced synthetically from β-pinene (a turpentine isolate) by the Glidden synthesis. This involves hydration of myrcene produced by pyrolysis of the pinene. Geraniol is found naturally in oil of citronella from which it can be obtained by distillation, and is widely used in floral perfumes together with a number of its esters. Another material which can be isolated from oil of citronella is citronellal, a terpene aldehyde with a lemon smell. Citronellol, which consists of two

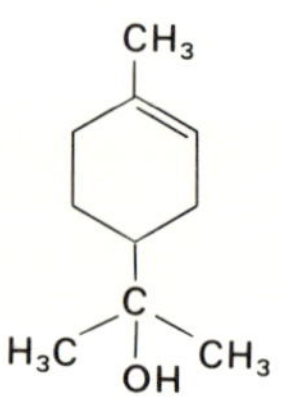

α-terpineol

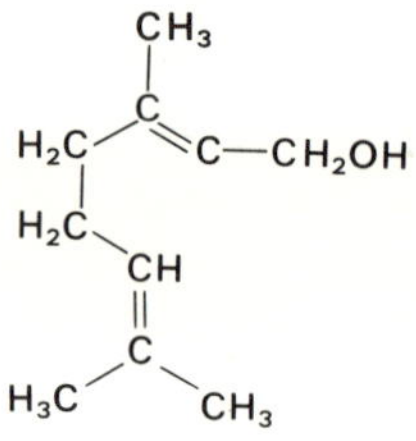

geraniol

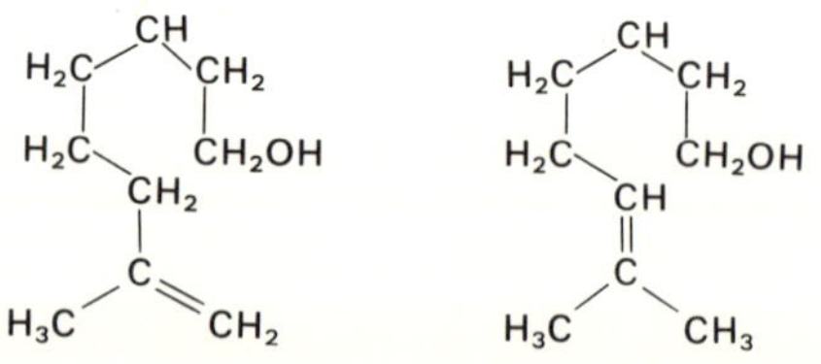

isomeric forms of citronellol

isomeric terpene alcohols, is prepared commercially by reduction of citronellal, and together with the acetate ester is also used in perfumery, especially in the production of synthetic attar of roses. Although the saturated aliphatic alcohols do not have pronounced odours, their esters are widely used for both perfumes and flavouring agents. Thus ethyl propionate and caproate have odours resembling pineapple, iso-amyl valeriate has an apple flavour and iso-butyl propionate smells of rum. Iso-amyl acetate is well known for its traditional role in flavouring 'pear drops', ethyl acetate has also been used for flavouring sweets and preparing fruit essences.

A number of tertiary alcohols such as phenylethyl methylethanol are also used in perfumes. These blend well and usually have desirable soft floral notes. Compounds of this type are usually termed carbinols and possess the tertiary alcohol grouping

$$R_2—\underset{\underset{R_3}{|}}{\overset{\overset{R_1}{\diagdown}}{C}}—OH$$

where R is —H, —⬡ or an alkyl group.

$$⬡—CH_2CH_2—\underset{\underset{C_2H_5}{|}}{\overset{\overset{CH_3}{|}}{C}}—OH$$

phenylethyl methylethyl carbinol

ALDEHYDES

The aldehydes often have powerful odours since the aldehydic radical —CHO is an odour-promoting group or osmophore. A disadvantage of the aldehydes as perfumery ingredients is their reactive nature which makes them prone to oxidation and polymerization. They are often purified commercially by making use of bisulphite addition compounds which can be decomposed to give a pure product.

An important terpene aldehyde is citral or geranial which occurs naturally in the essential oils of citrus fruits, eucalyptus and lemongrass. It has a powerful lemon smell and is widely used as a flavouring agent and in perfumery. The commercial product is an isomeric mixture of α- and β-citral obtained by the vacuum distillation of lemon-grass oil. It is used in the manufacture of ionone which is a structural isomer of irone, a violet-scented compound first isolated from the rhizome of the iris. Geranial can also be prepared by reducing citral. Citronellal is an optically active terpene which forms about 90% of the essential oil responsible for the odour of the Australian scented eucalyptus. As mentioned above it is used in the commercial preparation of citronellol.

citral   citronellol

The most important of the aromatic aldehydes is benzaldehyde, a colourless liquid with a powerful bitter almond smell which is prepared commercially from benzylidene chloride.

toluene   benzylidene   benzaldehyde
                   chloride

In a process operated by the Germans in World War II, the crude benzaldehyde is treated with soda ash and steam distilled. The residue containing sodium benzoate is used as a source of benzoic acid. Attempts have been made to oxidize toluene directly to benzaldehyde but none has achieved commercial importance.

In addition to its direct use in perfumery and flavours, benzaldehyde is an important starting material for the synthesis of benzyl benzoate, cinnamic aldehyde and a number of other perfumery intermediates.

The *ortho*-hydroxy derivative of benzaldehyde (salicylaldehyde) has a similar odour. It is usually prepared by oxidizing *o*-cresyl benzene sulphonate and then hydrolysing the product to give the free aldehyde.

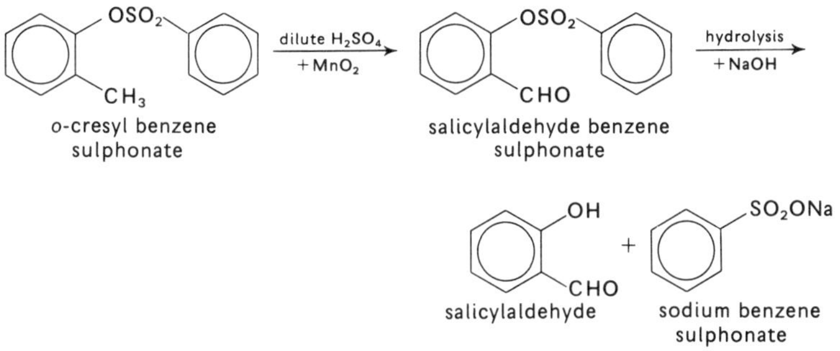

o-cresyl benzene
sulphonate

salicylaldehyde benzene
sulphonate

salicylaldehyde

sodium benzene
sulphonate

On refluxing salicylaldehyde with acetic anhydride in the presence of a catalyst such as anhydrous sodium acetate, coumarin is produced. This method of manufacture has now been replaced by a new synthetic route. *o*-Cresyl carbonate is prepared by condensing *o*-cresol with phosgene. Chlorination of the methyl side chain is followed by treatment with hot sodium acetate. Coumarin is produced by ring closure and separated from phenyl acetate and other by-products by fractional distillation.

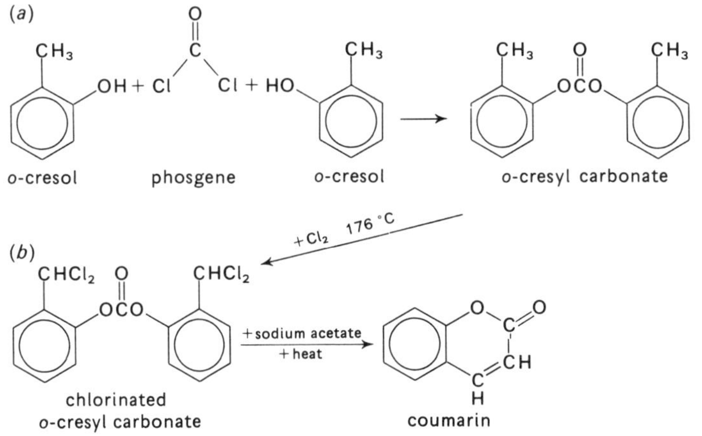

(a)

o-cresol   phosgene   o-cresol   o-cresyl carbonate

(b)

chlorinated
o-cresyl carbonate

coumarin

Pure coumarin has a smell of new mown hay and is of historical significance, being the first of the synthetic odiferous compounds to be prepared. Phenyl acetate formed during the synthesis is also a source of perfumery ingredients—ethylphenyl acetate having a pleasant honey smell. Coumarin is no longer used as a flavouring agent in Britain because of possible toxic effects.

Aniseed oil contains a high proportion of the phenolic ether anethole, which is used as a flavouring agent and soap scent. Oxidation of anethole is carried out on a commercial scale using an acidified solution of sodium dichromate to produce anisaldehyde. The reaction is 'stopped' at this point to prevent further oxidation to anisic acid by the use of sulphanilic acid as an inhibitor.

anethole → ($Na_2Cr_2O_7$ + dilute H_2SO_4) → anisaldehyde

During World War II the Germans developed an alternative method of producing anisaldehyde involving the oxidation of *p*-cresol methyl ether:

p-cresol → (dimethyl sulphate) → *p*-cresol methyl ether → ($KMnO_4$ + H_2SO_4) → anisaldehyde

Anisaldehyde has a pleasant smell which resembles hawthorn flowers and is an important ingredient of floral perfumes.

The pod of the vanilla plant contains from 1 to 2% of a delicately scented derivative of benzaldehyde, known as vanillin. This material has been used for many years to flavour custard powder, ice-cream and other confections, in addition to its use in perfumery. Most of the commercially used vanillin is now produced synthetically from iso-eugenol—the main constituent of oil of cloves—or from waste sulphite liquor obtained during the purification of wood pulp. The latter process is widely used in the USA.

The ethyl vanillin has an intensified flavour and smell and is also used to replace natural vanilla as a flavouring agent and perfume.

vanillin   ethyl vanillin

Another interesting derivative of benzaldehyde is piperonal, which is also known as heliotropin on account of its powerful odour of heliotrope. It is often used in the compounding of floral perfumes. Commercial production involves oxidation of safrole, one of the constituents of camphor oil.

By condensing benzaldehyde with acetaldehyde under slightly alkaline conditions at low temperature, cinnamaldehyde is produced. This has a spicy note and is the main constituent of cinnamon and cassia oil. Being unsaturated it is slowly oxidized in contact with air, but is used as a flavouring agent and perfume. Phenylacetaldehyde, which has a hyacinth smell, can be prepared from cinnamic acid.

piperonal   phenylacetaldehyde   cinnamaldehyde

KETONES

The most important members of this group are the ionones. Although they have a powerful odour of violets they are rarely found naturally and are never found in the flower. Commercially they are produced by condensing ketones with citral and ring-closing the resulting pseudo-ionones. They are widely used in floral perfumes such as violet, rose and lilac. β-ionone is also of importance in the synthesis of vitamin A.

A mixture of α- and β-ionones, known as 100% ionone, is available commercially and is produced by condensing acetone and citral, using dilute sodium hydroxide as a condensing agent. The resulting pseudo-ionone is then joined up to form a cyclic structure (ring-closure) using either concentrated sulphuric acid or phosphoric acid. The α- and β-ionones only differ in the position of their ring double bond, and can be separated if required by making use of the lower solubility of the α-isomer bisulphite addition compound.

By using methyl ethyl ketone instead of acetone the same synthetic route can be followed to prepare the four isomeric methyl ionones, which have a softer, pleasanter note than the ionones. The gamma isomer is preferred for use in perfumery.

The odiferous animal secretions musk and civet both owe their smell to the presence of the alicyclic ketones muscone and civetone. These ketones have unusually large ring structures (macrocyclic). As they are much in demand as fixatives and the natural products are very expensive, a great deal of effort has been made to produce synthetic substitutes. The so-called nitro-musks are inexpensive and widely used in place of natural musk. They are aromatic di-nitro and tri-nitro compounds with a strong musk odour. Musk xylene was the first of the group to be synthesized. It is made by attaching a tert-butyl side chain to *m*-xylene using a Friedel-Craft type reaction, and nitrating the product. Musk ketone is prepared in a similar way but the product is acetylated before nitration.

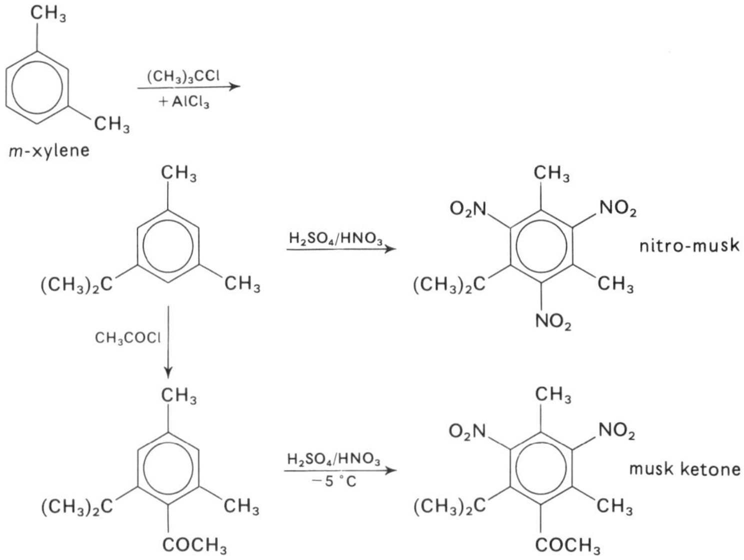

CH_3

m-xylene

$(CH_3)_3CCl$
$+ AlCl_3$

CH_3

$(CH_3)_2C$ CH_3

H_2SO_4/HNO_3

CH_3

O_2N NO_2

$(CH_3)_2C$ CH_3

NO_2

nitro-musk

CH_3COCl

CH_3

$(CH_3)_2C$ CH_3

$COCH_3$

H_2SO_4/HNO_3
$-5\,°C$

CH_3

O_2N NO_2

$(CH_3)_2C$ CH_3

$COCH_3$

musk ketone

Musk ambrette is derived from *m*-cresol by attaching a tert-butyl side chain to the methyl ether and nitrating the product with a mixture of acetic anhydride and nitric acid.

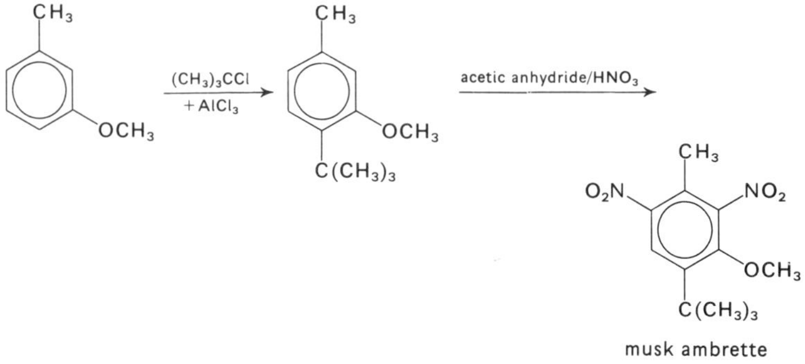

CH_3

OCH_3

$(CH_3)_3CCl$
$+ AlCl_3$

CH_3

OCH_3

$C(CH_3)_3$

acetic anhydride/HNO_3

CH_3

O_2N NO_2

OCH_3

$C(CH_3)_3$

musk ambrette

The ketone and ambrette are more expensive than the musk xylene but have a finer smell. In recent years other compounds having a pleasant musk odour have been evolved which are based upon an indene or naphthalene framework. These are cheaper and more stable than the nitro-musks.

Ružička has prepared a compound with a similar structure and odour to civetone which is known as 'Exaltone', but this has not attracted the commercial interest of the musks.

civetone   Exaltone

Piperitone is a fragrant ketonic terpene which is the main constituent of eucalyptus oils, and is found in certain Himalayan grasses. It has a peppermint smell and can be reduced to menthone or menthol and oxidized to thymol, the essential oil of thyme. Menthone is usually obtained from peppermint oil after the extraction of the menthol fraction by freezing, and also has a strong peppermint odour.

menthone   piperitone

One of the most familiar odiferous natural materials containing a ketonic group is camphor. Originally it was extracted as a white crystalline solid from the leaves, twigs and wood of the camphor tree, which is a native of Formosa. It has been in use for many centuries and is mentioned in the Koran and many early Arabian tracts. Synthetic camphor is now produced from α-pinene which is one of the fractions obtained by distilling turpentine. The synthetic product, however, is optically inactive ($\pm$ or dl-camphor), containing equal proportions of the d- and l-isomers, whereas natural camphor is dextro-rotatory.

The structure of camphor is interesting because it contains what is termed a *para*-bridged ring system. This can be thought of as a variation on the terpene skeleton produced by fusion of two isoprene nuclei.

camphor   camphor type nucleus   terpene type nucleus

Cineole is a terpene ether with a pronounced eucalyptus smell. It is found in the essential oils of wormseed and eucalyptus from which it is extracted by freezing and centrifuging.

COMPOSING AND BLENDING OF PERFUMES

The blending of odiferous materials to form a perfume is a skilled and lengthy process. The finished product must be economical, stable, and in a form in which it can be readily incorporated into commercial products such as soaps, talcs, creams and lotions. The perfume chemist must develop his sense of smell in order to recognize many hundreds of different odours or 'notes'. He 'composes' a perfume by mixing a variety of notes to form a smooth pleasing 'chord'. When working, he sits at a semicircular bench fitted with several rows of shelves. On these are many small bottles containing a vast array of essential oils and other odiferous substances, about 1500 in a full 'organ'.

Although much of the formulation of a new perfume depends upon

A perfumery research bench ('organ')

patient experimentation by the composer, there are certain fixed patterns of working to cut down wastage of time. In addition a variety of perfume bases are available as a kind of scaffolding upon which the skilled perfumer can build his own creation. For instance, a rose type floral perfume will usually have a foundation containing geraniol, citronellol, rhodinol and phenylethanol. These are ingredients of natural rose oil, they blend well and are readily miscible with other materials. Similarly many 'sophisticated' non-floral perfumes are prepared from a base of fatty aldehydes, such as lauric and methyl nonyl acetaldehydes, to which is added a spicy or woody top note.

Once the formulation of the perfume has been decided upon, the actual blending of the constituents is carried out in stainless steel or glass-lined containers using a propeller type stirrer. A low pressure steam heated jacket is usually provided to melt any solid ingredients, the temperature being progressively lowered as the more volatile components are added. Scrupulous cleanliness has to be observed at all stages of the blending process and the end-product stored in a cool dark place to avoid deterioration.

The formulation of perfume blends has been helped by an exhaustive classification of odiferous substances by Poucher, which occupied him for over four years. This is based upon an assessment of the volatility of a substance by noting the time of its complete evaporation from a paper slip. A scale of 1 to 100 is used. The most volatile components including groups 1–14 are called 'top notes', those in groups 15–60 'middle notes' and the rest 'base notes'. A typical perfume blend will contain volatile top notes such as the citrus oils for immediate impact, middle notes such as the ionones for body, and base notes such as musk or vanillin for persistence and balance.

Experimental blends are made by placing spots of single components on small strips of absorbent paper, and then clipping these together in different combinations to give various odour patterns. After smelling the chords produced in this manner, the papers are left for several hours and then smelled again to determine which of the

components are becoming dominant. Similar information can be obtained by rubbing a spot of perfume on the back of the hand and then following the course of the odour with time.

Apart from these short term changes in the nature of a perfume which are due to evaporation of the top notes, striking changes in character appear after periods of two or three months. A 'new' perfume has a 'rough' odour and often individual notes can be clearly distinguished. After ageing for a number of weeks, a much smoother and harmonious blend is apparent. This maturing effect is due to chemical reaction between the components of the product and is

12

analogous to the maturing of spirits. Apart from a lowering of volatility due to hydrogen bonding, the main changes are due to interaction of aldehydes with any alcohols or amines which may be present. This results in the slow formation of hemi-acetals and amine alcohols.

Recent evidence suggests that further reactions may occur such as the formation of Schiff's bases due to loss of water by the amine alcohols.

$$\overset{\displaystyle OH}{\underset{\text{amine alcohol}}{R\text{—}CHNH\text{—}R'}} \xrightarrow{\text{dehydration}} \underset{\text{Schiff's base}}{R\text{—}CH\text{=}N\text{—}R'}$$

Ester exchange has also been reported in the presence of suitable catalysts of the type

geranyl acetate		geranyl butyrate
+ benzyl butyrate	$\rightleftharpoons$	+ benzyl acetate

One of the difficulties in producing a satisfactory perfume lies in the sensitivity of some perfume structures to alterations in pH. This may cause a perfume to have quite a different odour on two persons with skins of differing natural acidity. A similar problem arose with the introduction of perfumed products in aerosol dispensers. In the presence of traces of moisture it appears that the propellant gas, trichlorofluoromethane, will react with the tinplate dispenser walls to produce small amounts of free hydrochloric acid. This has a damaging effect on the perfume. Another problem arising from the use of aerosol sprays is that when smelled in the form of a mist of fine droplets a perfume has quite a different character from that obtained on evaporation from a surface.

In the formulation of soaps, lipsticks, creams, lotions and shampoos, the main function of the perfume used is to mask the base fatty odour of the product. Care must be taken that the addition of the perfume does not discolour or upset the stability of the preparation. For instance terpineol and several other commonly used perfume components readily break down water-in-oil emulsions.

It is essential when composing a perfume, therefore, to bear in mind its probable role, and to carry out tests to ensure its satisfactory performance when incorporated in a specific product.

Oriental fruit flies trapped using a chemical attractant

TECHNICAL PERFUMES

Although the main demand for perfumes is in the preparation of cosmetics, there is now an ever growing demand for technical perfumes. These are frequently used to camouflage undesirable product smells. A classic example was the use of perfume to mask the unpleasant odour of the early types of permanent cold-waving lotion. Many other materials in common use have perfume added to make them more acceptable to the public. Examples are shoe and floor polishes, dyestuffs and printing inks, rubber, lubricating oil, tobacco, disinfectants and packaging tape. Disinfectant perfumed with pine and carnation has been used experimentally on the Metro in Paris, being sprayed from the rear of the trains.

The psychological impact of perfume has also been widely exploited commercially. Experiments carried out in supermarkets have shown that goods such as stockings sell far more rapidly if perfumed than if unperfumed. Laundries have also used perfume resembling the smell of new linen to spray laundered articles. Consumer research has shown that many people buy a particular branded article because they like its smell. With this in mind advertisements are sometimes printed with ink which is scented with the commodity concerned. Novelty greetings cards are also available which carry printed pictures of flowers impregnated with the appropriate floral perfume.

In the USA experimental systems such as 'Aromarama' have been tried in which appropriate odours are sprayed into the auditorium of a cinema during the action of a film to heighten the realism. During the last war odiferous materials were used in a number of interesting ways. A shark repelling substance was produced by the Americans for towing behind the life-rafts of airmen shot down in tropical waters. In addition some 15 000 tonnes of dimethyl phthalate were produced as an insect repellent for the use of the Allied armies fighting in the jungles of the Far East. Powerful perfumes were also sprayed from low flying aircraft over enemy occupied France as invisible and silent signals to the Maquis.

Farmers have prevented mice nibbling through the cords binding sheaves of wheat by soaking the cord in a liquid repellent to the mouse. Forestry officials in Denmark have used similar methods to prevent damage to the bark of young trees by hares. Odiferants have also been used for attractive purposes; surfaces coated with contact poisons such as DDT and 'Gammexane' are often treated with odiferous materials to attract insects.

Recently chemists in the US Department of Agriculture screened some 6000 chemicals in the hunt for odiferous substances which could be used to lure insects into traps. Some of these chemicals have proved highly effective at distances of nearly a kilometre ($\frac{1}{2}$ mile) and by combining them with an insecticide, large numbers of insects have been destroyed. The highly destructive Mediterranean fruit fly has been eradicated from the extensive Florida fruit orchards in this way.

An interesting example of the use of chemical attractants has been the synthesis of substances similar to those produced by the female insect to attract her male partner. Using these materials in traps such large numbers of male insects can be killed that the females remain unmated and lay sterile eggs. The first successful use of this technique in 1960 by the US Agricultural Research Service led to the elimination of the gypsy moth with a synthetic material termed 'Gyplure'.

Index

Oil-water boundary, 67
Oleates, 33
Oleic acid, 3, 4
Oleine, 53
Oleo, 34, 35
Oleochemicals, 54
Oleyl alcohol, 84
Olfactory recepter, 143
Olive, 2
Olive oil, 4, 19
Opacifying agents, 119
Open kettle rendering, 24
Optical bleach, 62
Orange, 2, 124
Orange oil, 154
Organ, 167
Organosol coating, 51
Osmophore, 159
Ouricuri wax, 96
Oxo alcohols, 88
Ozokerite, 99, 125

Paddle mixers, 95
Paint pigments, 46
Paint thickeners, 101
Paints, 44ff
Palm kernel, 2
Palm oil, 35, 135
Palm wax, 96
Palmitic acid, 3, 4
Pancake foundation cream, 107
Paraffin wax, 53, 99
Paraffin wax cracking, 85
Paste polishes, 94, 95
Patch testing, 139
Pear drops, 159
Pearl essence, 127
Pecans, 2
Pentaerythritol, 91
Peppermint, 148
Peptide linkage, 128
Percolation extraction, 22
Perfume composing, 167
Perfumery, 141ff
Pestle and mortar press, 17
Petroleum jelly, 126, 137
pH of soaps, 71
Phase volume fraction (PVF), 110
Phenyl acetate, 162
p-Phenylene diamine, 138

β-Phenylethanol, 157
Phenylethyl methyl ethanol, 159
Phosphates, 62
Phospholipids, 9
Photosynthesis, 10
Phytosterol, 13
Picramic acid, 138
Pine oil, 102, 105
Pine wood pulp, 44
α-Pinene, 166
β-Pinene, 158
Piperitone, 166
Piperonal, 163
Pitch liquor, 74
Pitching soap, 73
Plastic emulsion paint, 48
Plastic puffers, 107
Plasticity, 7
Plasticizer, 48, 49, 127
Plodders, 76
Pluronics, 93
Polishes, 56, 94
Polyester resins, 127
Polyethylene glycol, 117
Polyethylene glycol stearate, 100, 117
Polymethyl methacrylate, 51
Polymorphism, 4
Polyols, 91
Polythene ball applicator, 108
Polyunsaturates, 33
Polyurethane varnishes, 45, 52
Polyvinyl acetate (PVA), 48, 50
Polyvinyl chloride (PVC), 50
Polyvinyl pyrrolidene, 136
Pomades, 136, 153
Portable mixers, 113
'Pot life', 48
Powder compacts, 123
Powders, 121
Pregnendone acetate, 120
Preparation of emulsions, 112
Primary paints, 48
Printing ink, 43
Pristane, 9
iso-Propanol, 105
Propeller stirrers, 95, 113
Propyl gallate, 41
iso-Propyl myristate, 7, 126, 137
iso-Propyl palmitate, 7
Propylene glycol, 125